William Allan Macdonald

Humanitism

The Scientific Solution of the Social Problem

William Allan Macdonald

Humanitism
The Scientific Solution of the Social Problem

ISBN/EAN: 9783337365714

Printed in Europe, USA, Canada, Australia, Japan

Cover: Foto ©berggeist007 / pixelio.de

More available books at **www.hansebooks.com**

HUMANITISM:

THE SCIENTIFIC SOLUTION OF THE SOCIAL PROBLEM.

BY

W. A. MACDONALD.

LONDON:

TRÜBNER & CO., LUDGATE HILL.

1890.

TO

MELLOS [2048]

Tbis Book

IS

MOST HUMBLY AND WORSHIPFULLY

DEDICATED

BY

THE AUTHOR.

———

Those who to know the will of MELLOS strive,
  In these bold pages Truth divine will find ;
How man his tragic Limbo may survive,
  To praise the bleeding Saviour of mankind.

W. A. M.

PREFACE.

IN presenting this work to the public, no apology, I think, is necessary, for my conclusions do not interfere with the writings of other authorities on social questions. The material has been gathered from the ablest scientific authorities, but its method of application to the solution of the social problem is quite original.

The constantly diverging trains of modern thought necessitate some new and enlarged method of restoring unity and harmony; and I am baffled in all my efforts to find how this end can be attained through the theories of any one or more of the existing schools. In an inquiry which claims to be scientific, every sphere of human activity must be equitably adjusted, and the comprehensive nature of the undertaking may thus be easily realised. Had I framed together the mass of material which I have collected, the dimensions of this volume would have been multiplied at least threefold, which would have placed it beyond the reach of many whom it is intended specially to benefit. Accordingly I have written the work in language as popular as can be made consistent with an inquiry of this kind, and have sacrificed much for the sake of brevity. This course, however, does not detract from the value of the work as a message to scientists; for, although the matter may be well known to them, the inquiry points out the immense and largely untrodden field of scientific investigation connected with social problems.

The necessity for a scientific standard of morality, wherein all our actions may be reduced to numbers which have a definite relation to our moral destiny, has long been felt, and to this feature of my inquiry I desire to draw special attention.

W. A. M.

October 1889.

INTRODUCTION.

THAT there is a social problem to be solved may be pronounced an unwarrantable assumption. Some maintain that what is, is right, and man is therefore just what he ought to be; that, barring the conduct of a few malcontents who should be peremptorily suppressed, there is nothing to disquiet social order and harmony. These convictions, however, are not now shared by a majority of fair-minded people.

No end can be attained in the discussion of this question without first of all fixing man's moral destiny. This, again, may be viewed from two standpoints: (1) the destiny fixed by the perpetuity of the forces through which we are now controlled, and (2) the destiny fixed by any proposed changes in these forces. It does not alter the proposition to say that we, as the architects of our own fortunes, are the controlling agency; for, under all circumstances, every action or inaction begets one or more logical sequences.

In this inquiry it is my purpose to show that the controlling forces, or any modifications thereof which have been proposed, lead to the extinction of our race; and if this be our moral destiny, there is no social problem to be solved. I should, perhaps, have said the extinction of civilisation, instead of that of our race, for the extinction of all humanity is conditioned by the universal spread of civilisation, not of existing forms only, but also of all civilisations that may arise under the sway of any one or more of the existing schools of sociology. There being, therefore, no social problem to be solved in the abstract category, the only recourse is to shift the forces into the category wherein the perpetuity of civilised peoples is involved. I shall show that civilisations are built upon abstract theories, which have removed them from the natural laws of their development, and that a scientific basis for all our actions and activities must be obtained. This end can only be reached through unity and harmony of thought —an orderly restoration of the disorder and discord engendered by the ever-diverging schools of social metaphysics.

In speaking of the method of solving the social problem as being scientific, and of the founding of morality on a scientific basis, it is necessary to understand precisely in what sense the word science is employed, what is meant by the scientific method of inquiry, and how far science can be utilised in the solution of social problems.

In the same connection, the use of the word philosophy cannot be evaded, and the question may arise as to whether my inquiry is scientific or philosophic. Both science and philosophy have wide and narrow significations. It is the tendency of philosophy to move out of its restricted, discredited, abstract import, and make its sphere the unity and harmony of all the sciences, assigning to each its fitting place in relation to all the other branches, and to the totality of all natural phenomena. In this extended sense, there exists no such thing as philosophy—or, rather, it is still in its infancy —for such a scope would involve a solution of the social problem. In the following pages, I have embraced the whole sphere of human activity, by which my inquiry may be regarded as philosophic; but I have evaded this designation for two reasons : (1) it would create the impression that my method is based on abstract theories, and (2) philosophy has not won the distinction of making a synthesis of all the sciences. However, in my analytical chapters (PART I.), I have used the word philosophy in the abstract sense; and in the synthetical chapters (PART II.), I have used it in the sense in which I have termed my method as being scientific, believing that all inquiries of this scope and character should be regarded as philosophic, and that scientific investigations should have a more restricted sphere.

It now remains for me to show in what sense science and scientific methods may be employed in the solution of the social problem. In the first place, the conditions must all

be taken into consideration, which has not been done in any
of the investigations into the social problem; for humanity
and human destiny have been invariably eschewed, and these
are the most important conditions. For the purposes of this
inquiry, all knowledge may be derived from three sources.
(1) Thinking or reasoning about ideas. (2) Reasoning about
external objects which we behold around us. (3) Analysing
—resolving into their component parts—the said objects, ar-
ranging and classifying them, and deducing laws, principles,
relations, and theories therefrom. Setting aside the ideas
embodied in the Oriental Theocracies, realism, or reasoning
about sensuous objects, was the original method of acquiring
knowledge, and we find it distinctly marked in the early
Ionic philosophy; or, it should be said that Thales, the first
of this school, was a physicist rather than a philosopher,
although, in deference to the writers of the history of philosophy, I have included this and similar schools in my review.
This method of acquiring knowledge, namely, by ordinary
observation of sensuous objects—and these may also be the
rudiments of classification—may, however, be eliminated, for
it is the basis of science, and the Ionic physicists were
scientists without a science. In its restricted sense, philosophy is therefore confined to thinking or reasoning about
ideas, although it would be more correct to say that this is
the basis, for our ideas bear some relation to sensuous objects.
In philosophy, the basis is the idea; in science, the basis is
the object from which we obtain the idea. Science implies
knowledge derived from observation of and experimentation
with concrete or sensuous objects, including their classification,
laws, relations, &c. Any source of knowledge may be received
into the scientific category when it is placed upon an experimental basis. A scientific truth is obtained when all the conditions affecting the experiment have been observed, so that
a repetition always leads to the same results; but strong
evidence of truth may be predicated before this stage of perfection is reached, there being sometimes flaws in the conducting of experiments, due to unheeded accidents, or the lack of
skill and caution. The scientist cannot ignore the theories
of ideas promulgated by the abstract thinkers; for, holding

that knowledge is derived from experience, he must submit that ideas cannot transcend this limit—cannot attain theories which are not based on experience—so that all things which fall within this range may belong to the domain of science in the widest sense of the word. The existence of conditions beyond the domain of experience, if they variously affected the experiment, would be death to all scientific inquiry.

It may be urged, however, that the scientist cannot reduce all social phenomena to an experimental basis, that sociology involves history, which is not susceptible of experimental inquiry. This objection is unfounded, for history is an experiment in constant operation, and if the past has not led to results desired, he may suggest another basis, and show that history, like all other experiments, will repeat itself under the same conditions. He is permitted to apply the same logical methods to history as to other experimental inquiries. If it be, for example, a religious question, he may investigate the origin and development of the thousand and one religions of existing tribes, and compare them with those of civilised nations, inquiring at the same time into the origin and development of the languages, the social and political ideas, customs, traditions, ceremonies, the development of the bodily structures as obtained from geological or pre-historic records, &c. If the inquiry be a given historical incident, he may examine if the historian has delineated a violation of any natural laws, and, if so, whether the cause appears to be wilful misrepresentation or blind adherence to the superstitions of the age. The time is past when history is regarded as a dry record of battles and royal prerogatives, and the scientist has become the chief functionary in historical inquiry.

Although my dependence is mainly upon the physical sciences, yet I have not neglected mathematical methods, for in them is forcibly exemplified how unity and harmony of thought can be attained. The axioms appeal to every rational mind, and many unerring truths have been derived therefrom. If I have erred in naming my social assumptions axioms, my apology is, that if mathematical axioms became a party question, they would not appeal to us all as self-evident truths. In arithmetic, however, such unity and harmony do not exist, and this exception only proves that the mathematician and the scientist unite in striving after simplicity of method, and in forcing principles to their logical conclusions. The radix of our arithmetical system is 10, and many mathematicians contend that it should be 12; or, rather, that two new numbers should be inserted between 9 and 10 (12), making in all 12 digits instead of 10; and it has even been suggested that some power of 2 should be the radix. The reason why 10 is regarded as a false radix is because it has (besides unity) only two integral divisions, while 12 has four. It is plain that this question is capable of experimental inquiry, and in this sense it is scientific. The necessity for unity and harmony of thought, as well as simplicity of calculation, is aptly instanced by the metric system—although it is still imperfect—all measures of length, area, capacity, and weight being based on the length of a quadrant of the meridian measured between the equator and the pole, and they are all decimally related to each other. It thus appears that there are logical conclusions which enforce harmony of thought; and the only obstacle to their general adoption is (in the above instances) the lack of (mathematical) knowledge in the electorate. In my inquiry into the social problem, I have followed both of these mathematical methods, namely: (1) the founding of axioms, and (2) the forcing of scientific truths to their logical conclusions.

Another method in the attainment of unity and harmony consists in the solution of dualisms. In philosophy, for example, we have mind arrayed against matter; in metaphysics, the human soul opposes lower animal life, and in theology good and evil abound. Besides these, there are dualisms which have not yet been discovered, and my inquiry will consist largely in the discovery and solution of dualisms. There is an ultimate dualism which embraces all the interior and inferior ones, and a solution of the former involves a settlement of all the others.

In the discussion of some of the abstract theories, I have placed little reliance upon positive methods of proof, but have pointed out the weaknesses by showing how the authorities in the various schools contradict and abuse each other, thus necessitating a new and enlarged basis of inquiry; while in other instances I have cut short the negative proofs, and have depended mainly upon positive methods of demonstration.

My definition of the character of the scientist is deduced from a principle laid down by the first great scientist, Aristotle, who said that the object of inquiry into the nature of virtue was not to know what virtue was, but to be virtuous. In like manner, the true scientist does not inquire into other objects for the mere curiosity of knowing, but for the purpose of improving his conduct and that of his fellows—a man who has also the moral courage to bring all his feelings, thoughts, and actions in subordination to truth, despite the derision of men who are under the restraint of other forces.

The name of my system of social science, *Humanitism*, I have chosen for several reasons. The subject is one in which humanity stands out supreme, not only when referred to my social sympathies, my faith, and my hopes, but also on the ground of natural law. It is the ascendant force in our social system—the most potent agency in the social experiment. Other schools parade their practicality because they deal with man as he is. On this basis, it may be legitimately demanded that he be kept as he is and not be permitted to grow worse. In Humanitism, he is taken as he is and led to what he ought to be—a condition of ever-increasing health and strength, moral, intellectual, and physical, as well as ever-increasing virtue and happiness.

CONTENTS.

CHAPTER III.

POLITICAL ECONOMY.

CHAPTER IV.

THE THEORIES OF POPULATION.

CHAPTER V.

THE THEORIES OF RIGHTS AND TITLES.

CHAPTER VI.

ECONOMIC AGRICULTURE.

CHAPTER VII.

THE THEORIES OF THE STATE.

CHAPTER VIII.

STANDARDS OF VALUE.

CHAPTER IX.

RECAPITULATION AND COMMENTATION.

Part II.

THE REMEDY.

CHAPTER I.

THE EVOLUTION OF MAN AND NATURE.

CHAPTER V.

THE UNKNOWN GOD.

CHAPTER VI.

MATTER AND FORCE.

CHAPTER VII.

CONCLUDING NOTES AND COMMENTS.

PART I.

THE FAILURE OF ABSTRACT METHODS.

HUMANITISM.

PART I.
THE FAILURE OF ABSTRACT METHODS.

CHAPTER I.

PHILOSOPHY.

IN this chapter I purpose to give a few leading maxims of distinguished philosophers in all ages, and then inquire into the methods of the most famous schools of philosophy.

Cassendi and Mandeville regarded right as resting in the gratification of self-interest. Hobbes taught that self-preservation was the supreme good, and death the supreme evil; that right was what the sovereign commanded, wrong what he forbade; that war was a natural law—war of all against all, which could only be averted by the sovereignty of the strongest. Representing the Positivists, Mr. Frederic Harrison says : "To live for others and not for self is the real happiness of each of us, as it is our plain and simple duty." "The moral is the truly useful: it harmonises with the better part of human nature; for nothing can be useful which is unjust, for this is a destroying element in human society."—*Cicero.* "The corruption of the best things is the worst."—*Ancient Adage.* "Man is the measure of all things."—*Protagoras.* "The end justifies the means."—*Macchiavelli and the Jesuits.* "The individual is above society."—*Rousseau.* "Right is the authority to promote that which is necessary to fulfil one's destiny."—*Krause.* "Man has no rights except the right to do his duty."—*Comte.* "Act according to your conscience." — *Fichte.* "Know thyself."— *Socrates.* " The happiness of the individual is the sovereign principle."— *Basedow.* Moses Mendelssohn regarded the perfection of the individual as the supreme good and the source of moral conduct. Helvetius taught that the gratification of the sensuous desires was the first principle of morality, and that self-interest was the lever of all intellectual action. Epicurus also preached the sovereignty of the sensuous desires, but in the higher and nobler sense of the word, and was a champion of moderation ; while the opponents of these doctrines have preached abstinence from, or indifference to, objects which gratify our desires as the highest pleasure. The creed of moderation found an able exponent in Aristotle, and has had many disciples in China and Japan. Socrates taught that man should act from the intellect, not from the sense, and regarded self-denial, discretion, and especially a conviction of one's own ignorance and insufficiency as the basis of virtue. "Good is the aggregate of pleasure."—*Austin.* "Live harmoniously with nature."—*The Stoics.* "There exists no absolute virtue, and even so little an absolute vice : the only thing which deserves the name of virtue is moderation ; the only thing which can be named vice is the overstepping of moderation."—*Schroot.* Lycurgus, while believing that perfect physical development was the great essential in the attainment of human felicity, inculcated, in his Spartan Laws, patriotism and the vanishing of the individual into society as the highest virtue ; and these Socialistic doctrines also found favour with Solon, Plato, and Aristotle—also with the rulers of the Roman Empire ; which principle, however, degenerated to a despotic extreme under the Emperors. The Pythagoreans were the first to recognise the union of right and virtue in the community of property, and Plato carried this Communistic theory so far that he advocated the pairing of men and women by State authority in such a manner as to breed the best offspring. Apart from modern Socialism, there are numerous philosophers who maintain that ignorance, disease, poverty, misery, and immorality go hand in hand with the extreme struggles for existence, and vindicate the restoration to virtue and happiness through amelioration of the material condition of the masses by State measures. Herbert Spencer robs the State of all its functions excepting the protection of life, liberty, and property, and the administration of justice ; and his theories are ardently defended by a large and influential organisation of Individualists. "The liberty of man consists solely in this, that he obeys the laws of nature, because he has himself recognised them as such, and not because they have been imposed upon him externally by any foreign will whatever, human or divine, collective or individual."— *Bakunin.* "Right is the outer compulsion to act, or cease to act, for the purpose of respecting the liberty of others ; and morality is the inner necessity to act, or cease to act, without regard to external motives."—*Fichte.* "The object of life is the welfare of the individual and the community : the individual, even the reigning sovereign, must subordinate himself to the general weal."—*Frederick II.* The Nihilists, whose doctrines were first taught amongst the pious in ancient India, taught that the highest aim was the transition of the individual into the all-god : all earthly existence was an evil, and the highest happiness reposed in non-existence. This doctrine has found many adherents in all ages, and their supporters at the present day are called pessimists. "Act so that the maxim of your will may be capable of being regarded as a principle of universal validity, or so that from the thought of your maxim as a law universally obeyed no contradiction results."—*Kant.* A school of modern philosophers, under Locke, Bentham, Mill, Bain, and others, have attempted to revive the ancient rule, namely, the greatest happiness to the greatest number, and many living philosophers have adopted it as the supreme law of their conduct.

The above brief summary is sufficient to prove the absence of harmony amongst our philosophers during the past two or three thousand years, and I am convinced that no apology will be required for seeking another method of attaining unity and concord ; not that these precepts are to be despised, for there are many golden rules amongst them. It will have been observed that the predominating ends to be attained are virtue, happiness, knowledge, right, the supreme good,

morality, usefulness, justice, moderation, destiny, fulfilment
of our desires, harmony with nature and conscience, perfection, self-denial, discretion, liberty, development. Even if
our philosophers were a unit in the acceptance of any one of
these ends, no solution would yet be possible, for discord
would at once arise in points of definition. What is moderation, right, justice, perfection, liberty, &c.? The answers to
these questions are the great disturbers of the peace.

However, lest there should be some dominant school which
bids fair to arrive at some universally recognised truth, let us
take a brief historical survey of philosophy, ancient, mediæval,
and modern. Were we to examine the history of ancient Greece
in detail, little would be left unsaid in philosophy, despite the
imperfect fragments of their literature which have been handed
down to us; for ancient and modern problems are essentially
the same, none of which have yet been solved, and modern philosophy has added nothing to guide us in our conduct or enable
us to attain a clearer conception of a First Cause. The ruling
thought during the Middle Ages was a theology rather than
a philosophy, which will be considered in a separate chapter.

Lewes, in his "History of Philosophy," divides ancient
philosophy into nine epochs, each possessing the following
characteristic:—

"1. Philosophy separates itself from theology, and attempts
a reasoned explanation of cosmical phenomena."—*Thales to
Zeno.*

"2. The failure of cosmological speculations directs the efforts
of philosophy to the psychological problems of the origin and
limits of knowledge."—*Heraclitus to Democritus.*

"3. The first crisis. The insufficiency of philosophy to solve
the problem of existence and to establish a basis of certitude
produces a sceptical indifference."—*The Sophists.*

"4. Philosophy emerges from the crisis by a new development of method—the application of dialectics as a negative
process preparatory to the positive foundation of inductive
inquiry."—*Socrates.*

"5. Development of ethics consequent on the Socratic cir-
cumscription of the aims of philosophy."—*The Megarics to the
Cynics.*

"6. Restoration of philosophy to its widest aims. Attempts
to follow up the negative dialectics of Socrates with an

affirmative solution of the chief problems. The necessity for a criterion of philosophy becomes for the first time distinctly recognised. The answer to this question gives a logical basis to the subjective method." [1]—*Plato*.

"7. Philosophy for the first time assumes the systematic form of a body of doctrine, all its conclusions respecting existence being referred to principles of logic. The criterion stated by Plato is systematised and applied by Aristotle. A method of proof takes its place amongst the chief instruments of thought."—*Aristotle*.

"8. Second crisis in philosophy. The radical imperfection of the subjective method again becomes manifest in the impossibility of applying its criterion."—*The Sceptics to the New Academy*.

"9. Reason allies itself with faith, and philosophy renounces its independence, becoming once more an instrument of theology."—*Neo-Platonism to the End of Ancient Philosophy*.

In the pre-Socratic philosophy, beginning with *Thales* (640–550 B.C.), the inquiry was for an ultimate principle by means of which nature could be explained.

That philosopher, placing reason above sense, arrived abstractly at the conclusion that water was the source of all things. Amongst his contemporaries, however, one concluded that air, another fire, and another a chaotic primeval matter was the ground of nature. The Pythagoreans attempted a higher solution by tracing the essence of things to numbers—proportion and harmony; but it is disputed whether this was regarded as a material or an ideal principle. The Electics, driving abstraction to the ultimate, enunciated the formula that "only being is, and non-being (becoming) is not at all." This theory of existence, especially in the minds of contemporary philosophers, was essentially monistic, not dualistic. Heraclitus, regarding Electicism as dualistic, sought to harmonise this theory of pure and phenomenal being—the dualism of the one and the many—by his doctrine of "eternal flux," thus resolving the being and the non-being into both—becoming; the world was therefore neither the one nor the other, because it was both at the same time. Empedocles and the Atomists endeavoured to combine the principles of the Electic and Heraclitic schools in another way. The

---

[1] Lewes uses the "subjective method" in the sense in which I use the word "abstract," and "objective" in the sense in which I use the word "scientific."

former considered the four elements to be eternal, imperishable, and self-subsistent, but divisible, and mingled and
moulded by two moving forces—friendship the uniting and
strife the disuniting force. Friendship first held these elements
in unity till strife came to produce discord. The Atomists
derived phenomena from an infinite number of primeval atoms
which were all alike in quality but unlike in quantity. They
were material particles, eternal, immutable, and indivisible,
endowed with a primary motion which gave form to natural
phenomena, and which were also the source of life.

This was essentially the character of Greek philosophy
when *Anaxagoras* (born about 500 B.C.) appeared to give a
new turn to philosophic thought. The philosophy of his day
was becoming very complicated, and a new basis was eagerly
sought.

This philosopher perceived that the current conception of
moving forces was vague and imperfect, and lacked the
element of design in the explanation of the phenomena of
nature. Accordingly, he introduced a world-forming intelligence, a spontaneous, immaterial designer, free from, and
merely a mover of, matter. This gave an idealistic turn to
the materialistic and realistic tendency of the age, mind now
being, in the creed of this school, the ultimate principle of all
things—which doctrine, however, required yet to be demonstrated. First there were the primitive constituents of inert
matter, and then came the mind to put them in order.
Anaxagoras, by his ingenious method of explanation, restored
unity in philosophic thought, and closed the pre-Socratic
period of realism.

In the preceding philosophy, consciousness was made
subjective to objective realities, objectivity being the only
source of knowledge; but now, in the *Sophists*, a new era
dawned.

This school sought to establish the principle of subjectivity, drawn from the Heraclitic doctrine of eternal flux.
The individual had therefore now a higher existence than
objective phenomena, State laws, inherited customs, religious
traditions, and popular beliefs, thus giving reason supreme
sway. Although having no philosophic system, they reduced
knowledge to method—were an encyclopædia of knowledge.
They strived after notoriety, especially amongst the rich,
demanded stipend for their preaching, led a wandering life,

made great rhetorical display, and styled themselves professional thinkers. However, they made more use of their wit than their knowledge, believing that a knowledge of facts was not necessary in discussing the problems of the times. Whatever their failings may have been, they taught the people to think methodically and speak correctly. Protagoras, the first Sophist, was banished from Athens as a blasphemer, having remarked that he was unable to know whether the Gods existed or not—a precedent which might now be cited against the Columbus of Agnosticism.

The way having been paved by the Sophists, *Socrates* (469–399 B.C.) inaugurated the era of the philosophy of objective thought. He turned the mind upon itself, and endeavoured to discover the nature of virtue.

He believed in a supreme intelligence who was the cause of the marks of design, and defended the existence of the Gods, arguing from the structure of organised beings, and founding his reasoning on the general principle, that whatever existed for a use must be the work of intelligence. He felt inspired by the divinity to preach the gospel of his philosophy. For him there was nothing to be learned in the fields or amongst trees, and he accordingly confined his duties to the market-places and workshops, discussing practical problems of life, and convicting men of their ignorance, for which office he disdained all compensation, except the reward of a pure conscience. Genial, patriotic, and brave, he won many scholars, although he founded no school, and the sages of all times, agreeing with the Delphic oracle, have pronounced him the wisest of men. Having acted his principles, his biography is his philosophy. He was so pious that he did nothing without the sanction of the Gods, and disdained to become a politician. To know one's self and one's ignorance was the soul of his philosophy. Virtue was inseparable from knowledge, and could only be attained by the exercise of right reason. Man was by nature a sympathetic and reasoning animal, with strong self-preserving and self-asserting instincts, and did not do wrong intentionally, but through ignorance. The function of reason, which in Socrates was the source of his piety, was recognition and realisation of the truth. To know what was right was to do what was good. Happiness consisted in the free and unhindered exercise of man's characteristic function—reason. He recommended his doctrines in preference to those of other philosophers on account of their great utility and practical usefulness; and beauty and morality consisted in utility. Religious duties were reverence, gratitude, and obedience, deeds being more effective than words. Right and law were identical—both the laws of the State and the unwritten laws of the Gods, which were an universal sentiment. It was impossible to escape the unwritten law in the conscience of men, but human law could sometimes be violated. The voice of the people was the voice of the Gods. Virtue was a science, and was therefore knowable, teachable, and learnable. Happiness was the highest good. To want nothing was divine. A real knowledge of God and godly things was inaccessible to man. This great philosopher, having been accused by influential and unprincipled demagogues of disbelief in the Gods of the State, of introducing new Gods, and of corrupting the Athenian youths, was condemned; and, choosing death before exile as being the more consistent with his philosophy, he died a victim of misunderstanding, of fidelity to ancestral custom, and of reactionary thought.

A pupil of Socrates, Antisthenes, the founder of the *Cynics*, adopted and enunciated the general principles of his great master, but carried the doctrine of self-control and freedom from want to its extremity.

He taught that virtue was the supreme end of human life, and the only good. Enjoyment, sought for its own sake, was an evil. Virtue was sufficient for happiness, and its essence was self-control. The sage was indifferent to honour, riches, and marriage. No form of government, actual or thinkable, suited the cynic; for his happiness was bound up in isolation from society—he was a "citizen of the world." Virtue was the only true object of worship, and the religion of the State should be no more binding on the sage than were the laws of men.

The time had again arrived when a conciliation or fusion of the diverging schools of philosophy was needed, and the task fell to the lot of *Plato* (428-347 B.C.). This philosopher and his great successor, Aristotle, became the all-absorbing and all-dispensing luminaries of this period of Greek literature. The Socratic germ, as the saying is, is traced to the Platonic blossom and the Aristotelian fruit. Socrates was essentially the educator and man of action, Plato the man of literature, and Aristotle the man of science. Plato spent eight years with his teacher, Socrates, and Aristotle enjoyed the society of Plato for twenty years. Alexander the Great having been a pupil of Aristotle, as was also Pericles a pupil of Anaxagoras, philosophy must have exercised considerable influence in ancient politics, just as political economy sways the politics of our day.

The focus of Plato's philosophy was the theory of Ideas, which he conceived as being freed from time, space, and materiality; but the Idea, which was his basis, enjoyed a certain community with material objects, and to some extent dwelt in them. Especially from the highest of these Ideas, which he regarded as the efficient cause of concrete existences, Plato derived his conception of the Good, and then he personified his idea-world as Gods. The highest Idea, the Good, he identified with the supreme deity. The method of acquiring knowledge was through logic; and mathematical, being mid-way between philosophical and sensible cognition, mathematical objects held the mean between things sensuous and ideal. In his old age he attempted to reduce ideas to numbers, which marked another stage in the evolution towards realism. From such a foundation, then, he fashioned his theories of virtue, of human conduct, and of the State. His idea of the Good, as the aim of human activity, was rather a commingling of knowledge and pleasure than merely a life of both. The destiny of the soul was escape from inward and outward evils of sense, purification and emancipation from all corporeal influences, striving to become pure, just, and like God, suppression of sensuous delights, retirement into cognition of truth—wrapt up in philosophy. He repudiated the hedonistic theory that pure happiness was attainable, for it transformed itself so readily into pain. His theory of virtue was quite Socratic ; and, with Socrates, he believed the soul to be immortal. His theory of the State was the outcome of his philosophy. The supreme duty of the State was to train the citizens to virtue. The virtue of the rulers was wisdom, that of the warriors valour, and of tradesmen and manual labourers self-restraint and willing obedience. The individual was to be completely absorbed in the State, moral into political virtue, and the laws were to be immutable. There was to be but one family and one property proprietor—the State. The *I* was to glide into the *We*. Plato's philosophy did not lead him to mention the word benevolence ; and he regarded the natural condition to be the casting out of feeble and imperfect children, and the sick were neither to be tended uor nourished. His political institutions were decidedly of the aristocratic stamp, preference having been given to an unlimited monarchy, the ruler being a perfected philosopher, and all the politicians were to be able to philosophise fully and correctly. There was to be but one ruler, because there were so few philosophers.

In *Aristotle* (384–322 B.C.) we find the analyser, the practical man, and the philosopher combined. Unlike Socrates, he held cities and multitudes in contempt; and, unlike Plato, the love of facts and nature being deeply rooted in his being, he moved from the concrete to its ultimate principles, from the datum to the idea, his method therefore being inductive or scientific—analytical; Plato's was deductive or philosophic—synthetical.

Aristotle reduces his philosophy to four principles: form (essence), matter, moving (efficient) cause, and end, the first mentioned being substituted for the Idea as enunciated by Plato. He combated the Platonic doctrine that the Ideas existed apart from concrete substances. Matter completely deprived of form did not exist; mere matter was a purely abstract idea. But there existed an immaterial form-principle, a separate and independent existence, which was to be distinguished from the inseparable forms inherent in matter. Matter was passive, was the ultimate source of imperfection, but caused individuation in things. Motion (change) was the evolution of potentiality into reality; and as motion implied a moving cause, God, though himself unmoved, generated motion. He was the immaterial and eternal form, actuality without potentiality, a self-thinking reason or absolute spirit, absolutely perfect, universally loved, into whose image all things sought to gravitate. Aristotle's system was that of the Heraclitic becoming, which was a marked tendency towards the undermining of the Platonic dualism. He was not a pious worshipper of the polytheistic theology of his country. He granted the existence of morality, like that of all physical objects; and his inquiry into virtue was not to know what virtue was, but to be virtuous. All moral problems were to be abandoned which had no direct bearing on practical affairs. He disputed the Socratic teachableness of virtue. Moral principles were universal, belonging to all times, and not to any given age. His supreme Good was the purpose of our existence; and the chief end of man, as humanity and not as individuals, was to reason, through the exercise of which process he grew virtuous, simply by habit. As for the individual, he had only to recognise and obey. The test of virtue, right conduct, was an adjustment between two extremes—moderation; and it was through the exercise of man's distinctive faculty, reason, that he fulfilled his destiny, and attained happiness as a necessary consequence. The essence of pleasure was rational (virtuous) activity. A life devoted to sensual enjoyment was brutish, an ethico-political life was human, a scientific life divine. The soul, the animating principle, was related to the body as form to matter, and ceased with the body; but the (active) reason, the separate, immaterial, self-subsistent, and divine principle, was immortal. (Here may be traced a dualism.) Man, being a political animal, could neither attain virtue nor happiness in his individual capacity; the existence of the State was therefore required to assist, protect, and afford opportunity for the practice of virtue. The State was higher than the individual and the family; and, though originating for the protection of life, ought to exist for the promotion of morality and rectitude. Obedience without intelligence was slavish. Although himself an aristocrat and laying little stress on forms of government, believing that given times and given conditions were important factors, he preferred, in general, a constitutional monarchy—an harmonious combination of the monarchical, aristocratic, and democratic elements—and objected to oligarchy on the ground of happiness to the few, not to the many. A good government was one in which the rulers sought the public good; bad, when they preferred their own interests.

Although with Aristotle the period of Greek philosophy virtually closes, yet it is necessary for the purposes of our inquiry to notice briefly the leading schools of succeeding generations in order to understand how philosophy merges into Christianity. Up to this period, we have first the rule of realism, then of idealism, which attained its height in Plato; and with him dawns also again the realistic phase, which culminated in Aristotle, who, in his turn, gave birth to the empiric school of modern philosophy.

The chief feature of the post-Aristotelian philosophy was *Stoicism*, the founder of the Stoic school being Zeno.

Their doctrines were essentially Socratic. All knowledge originated in experience or sensuous perception, the soul at first resembling a piece of blank paper on which impressions were yet to be made. God and matter being one substance, the Stoics united pantheism with materialism. Nothing incorporeal existed. The immediate effect of this union of God and the world was the subjugation of the dualism of the spiritual and the material. Matter in itself was motionless and formless; force was the active, moving, and forming agent—God. The world was God's body; God the world's soul. This sensualism of the Stoics, however, stood out in bold and strange contrast with the spiritualistic tendency of their moral conduct. Their moral principle was to live a life in harmony with nature. Individual considerations were subjected to those pertaining to the whole. Virtue, the sole end of rational beings, was also their sole happiness and sole good. The virtuous were not disturbed by want or the loss of earthly possessions, and were even indifferent to health. These were no evils; only vice was an evil, because it was contrary to nature. Amongst the later Stoics virtue consisted in absolute reason, absolute justice, absolute self-control over pain, and absolute mastery over lust and desire. The sage knew all that was to be known, having a true perception of the nature of things, divine and human, and did not lean upon the political element for his morality. The State emanated from man's social instincts, and it was a contradiction of nature that men should be divided into hostile tribes; the whole race should form one community under one political code. At the end, all things were reabsorbed into the deity, the universe being first resolved into fire, and then a new evolution began, and so on in eternal succession, and the soul endured till the end of the world-period in which it existed.

Epicurus (341–270 B.C.), the founder of *Epicureanism*, sought the supreme Good in human happiness, which was nothing but pleasure.

Virtue had no value in itself except so far as it afforded an agreeable life. Pleasure was tranquil satisfaction enduring throughout the entire life, not momentary, and true pleasure could be attained by calculation and reflection. Those pleasures which prepared us for pain should be rejected, and those pains accepted which prepared us for greater pleasures. Spiritual joys and sorrows were superior to those of a fleshly nature, the latter being temporary. Pleasure demanded sobriety, temperance, limited wants, and a life in accord with nature. Above all evils not to be dreaded was death, which no wise man feared. Not to live was no evil.

All phenomena were natural and intelligible, the interference of the Gods being unnecessary. The Gods had human form, but were without human wants, which was the highest ideal of happiness; they dwelt between the spheres, and did not concern themselves with human affairs. Atoms and space were eternal. The soul was material, was composed of infinitely small atoms, and was closely allied to air and fire. Sensuous perception depended upon material images.

The divergence of thought having again led to despair, an era of *Scepticism* was the result.

The Sceptics denied the certainty of knowledge, denied also that they were certain that they knew nothing, and based their happiness on the suspense of judgment. The sceptical mood and peace of mind was a condition of pure indifference —freedom from all care and desire, freedom from all knowledge of good and evil. Truth could neither be attained by man's perceptions nor by his ideas. All objective knowledge was impossible, and all that could be done was to follow the course of probability. There was no criterion of truth.

This Scepticism continued until the total decline of Greek philosophy, and after the Christian era. The Romans, whose Empire now held sway, conquered the Greeks, but were conquered intellectually by Greek philosophy, which now formed a constituent element in Roman culture, the Roman mind being barren of philosophic originality.

Alexandria had now become the great centre of culture, as well as of commerce. This great Egyptian city had also now become a political unit, and gathered men of letters from all civilised nations, notably Greeks and Jews.

The dualism of subjectivity and objectivity, as well as of matter and force, still remained unsolved, and the last attempted solution ended in *Neo-Platonism*.

The essential characteristics of the Neo-Platonic philosophers, if their system can be called a philosophy, were enthusiasm, magic—a pretended intercourse with spiritual beings ; most of them were addicted to sorcery, and the more notorious professed to be able to foreshadow future events and to work miracles. According to the most conspicuous representative of this school, Plotinus, a knowledge of the truth could not be attained by proof, but through visions, ecstasy, divine rapture

—a swooning into God. But this condition to which Scepticism had led failed in its object; for the Sceptics had striven after absolute peace, while this trancing into the Absolute proved the source of unsatisfying restlessness. The process produced a monism indeed, the philosophic dualism now being solved; but, by outwitting reason, proved the death of philosophy; so ancient philosophy became annihilated in the struggles to subjugate a dualism, and the death of philosophy was the birth of Christian theology.

Various attempts had afterwards been made to revive ancient philosophy, and especially, after the eleventh century, there developed a sort of Christian philosophy—*Scholasticism*. Its function was a conciliation between Church dogma and philosophic thought—between faith and reason; an attempt to rationalise revelation. So violent a dualism was destined to a lamentable fate, and terminated in a temporary victory of faith. Gradually, however, philosophy shook off the fetters of dogma, and asserted independent sway. Scholasticism was nurtured mainly by the philosophy of Aristotle, as then understood. The Church became a sink of corruption, its authority began to decline, and there arose in the Reformation a struggle for independence in religious thought. Classical literature revived, the natural sciences began to flourish, and the mind showed a tendency to penetrate more into nature than into itself.

· Lewes divides modern philosophy into eleven epochs, the transition period marking the struggles of philosophy to emancipate itself from theology, which it accomplished at the close of the Middle Ages. Ancient philosophy flourished during a period of about a thousand years before its capture by theology; and a similar period of time elapsed from the extinction of the Greek schools to the period which marks the separation of philosophy from theology in the sixteenth century. These epochs are as follows :—

" 1. Philosophy again separates itself from theology, and seeks the aid of science."—*Bacon, Descartes.*

" 2. The subjective method carried to its extreme results in pantheistic Idealism."—*Spinoza.*

" 3. Philosophy pauses to ascertain the scope and limits of the human mind."—*Hobbs, Locke, Leibnitz.*

" 4. The problem of an external world discussed on psychological data."—*Berkeley.*

" 5. The arguments of Idealism carried out into Scepticism. —*Hume.*

" 6. Attempts to discover the mechanism of psychological action ; the Sensational school."—*Condillac, Hartley.*

" 7. Second crisis : Idealism, Scepticism, and Sensationalism producing the reaction of common sense."—*Reid.*

" 8. Psychology finally recognised as a branch of biology—the Phrenological hypothesis."—*Gall.*

" 9. Recurrence to the fundamental question respecting the origin of knowledge."—*Kant.*

" 10. Philosophy once more asserts a claim to absolute knowledge."—*Fichte, Schelling, Hegel.*

" 11. Foundation of Positive Philosophy."—*Comte.*

Of these epochs, the eighth and eleventh do not belong to this chapter, for these belong to science rather than philosophy, and have nothing in common with abstract ideas. The Positive Philosophy marks the tendency of science to capture both philosophy and religion.

The way having been prepared by Bacon, *Descartes* (1596-1650), a French scientist and soldier, is commonly regarded as the father of modern philosophy, and, having repudiated all previous systems, has received credit for a good deal of originality. His standpoint was first to expel from the mind all inherited customs, and to doubt everything which had the least appearance of uncertainty.

The existence of all objects of sense was to be doubted, our senses often proving deceptive, and even the truths of mathematics ; for we could not know whether such was intended for finite beings, whether God had created us for mere opinion and error. But one thing we could not doubt, namely, that we who thought existed. "I know, therefore I am." From this certainty followed the whole nature of spirit. We could think ourselves all away except our thought. We were therefore essentially thinking beings—spirit, soul, reason, intelligence. It was impossible for us to think and not to be. Our ideas were partly innate, partly received from without, and partly formed by ourselves. Amongst all of these, we conceived God to be the predominating idea, which could only be implanted in us by a perfect being—an actually existent, infinite God, more perfect than ourselves, and could not be known by abstraction and negation, or through the senses; it must be innate, and God's existence must be inferred from our own imperfections, particularly from our knowledge thereof. We knew through the innate idea and God's necessary attributes that he possessed veracity, and it would be a contradiction for him to deceive us or cause us to err. Were the ability to deceive a. proof of superiority, the will to do so would be a proof of wickedness. All certainty emanated from God, which was the source of natural philosophy, or the theory of the duality of substance, requiring nothing else for its subsistence. All error arose from misuse of freedom of will. God had the source of his existence in himself, and was the cause of himself; but created substances, mind and matter, only required for their existence God's co-operation. Spirit and body possessed nothing in common; and as matter was essentially extension, and spirit essentially thought, neither of them having anything in common, the union of soul and body was mechanical—a forced union of essentially opposing natures. The seat of the soul and the thoughts was not the whole brain, but the pineal gland, which had no duplicate like other parts. The lower animals, having no pineal gland, self-consciousness, or thought, were mere machines.

Descartes having failed to reconcile his dualism except through a forced, one-sided, and external agency, the main difficulties observed amongst his successors were that, as matter and thought were different and opposed, bodily affections could not consistently act on the soul, or the volitions of the soul on the body, and that the soul could not attain a knowledge of things corporeal and external—the extended things.

The next philosopher of note who elaborated the doctrine of Descartes was *Spinoza* (1632–1677), a theologian of Jewish extraction. The basis of his philosophy was three notions, namely, substance, attribute, and mode.

There existed only one substance, and it required nothing for its support—a self-conceivable and unlimited unit—God, of which we cognised two attributes, thought and extension. It was inconsistent to think of the finite without the infinite. The absolute unconditioned substance was the cause of every finite existence, and was itself all being. God's freedom consisted in the inner necessity of his working nature. There was nothing bad in God, and nothing happened contrary to his will. The good was the useful, which enabled us to procure greater reality, and to preserve and promote our being—reason. The highest knowledge was the knowledge of God; the highest virtue of the soul to know and love him. From this knowledge we derived happiness, joy, and peace, and deliverance from discord and discontent. Happiness was virtue itself, not virtue's reward.

Spinoza's philosophy is a transformation of the Cartesian dualism into pantheism, and his God is identical with nature and natural law—an inner working necessity or persistent force, which admitted no free-will on the part of finite beings. In applying his philosophy to human conduct, he regards evil as only relative, not positive—a negation which appeared positive to finite minds. Descartes' isolation of thought and extension, spirit and matter, is now brought into unity of substance by Spinoza, but they are not a unit in themselves, being indifferent to substance itself, and so the separation is still complete. The position now is either to explain the ideal from the material side, or the material from the ideal side. Both of these attempts have been made, the result being that philosophic thought from that day to this has been divided into the schools of *Idealism* and *Realism*. The latter is also known under the names of empiricism, sensualism, and materialism.

The way having been cleared by Hobbes, the real founder of the modern Realistic school is *John Locke* (1632-1704). His theory of knowledge may be derived from the empirical interpretation of Aristotle, and has already been noticed under Stoicism, namely—

That there were no innate ideas in the mind, that all knowledge originated in experience. The understanding (soul) was at first a blank page. Experience was twofold: (1) the perception of external objects—sensation; and (2) the internal operations of the soul—reflection. The understanding derived all its ideas from sensation and reflection. His moral principle was happiness. From his borrowed postulate, namely, that nothing exists in the mind which has not had a prior existence in the external world, the inference has been drawn that he gave matter precedence to mind, and made the soul a material substance.

The empiricism of Locke was followed by the scepticism of *Hume*, who derived the doctrine that thought (soul) was merely a succession of transitory ideas, was therefore an illusion, and must necessarily cease with the movements of the body. But such logic, not being congenial to the English character, was driven to its ultimate by the French philosophers.

Condillac, commencing with Locke's postulate, resolved sensation and reflection into one condition—sensation. He did not deny the existence of God, and maintained the immortality of the soul. It was reserved for *Helvetius* (1715–1771) to draw the logical conclusion of sensationalism, basing his moral principle on the gratification of the sensuous, which aroused the wrath of the priesthood. Self-interest, in his philosophy, was the mainspring of all our actions, and self-love (self-interest) led to a life of bodily enjoyment only. It was irrational to expect men to do good merely for the sake of good, or commit evil for the sake of evil. In order to become an active agent, morality must be sought in this empirical source, and legislation should shape its measures accordingly.

Before the Revolution in France general demoralisation reigned supreme. The State had degenerated into a despotism and the Church into a tyrannical hierarchy. Unlike England, France became specially receptive of sensualistic ideas. It was against the interests of the ruling classes in England that the influence of the Church should wane; and the Scottish philosophy was aimed at the empiricism of Locke and the scepticism of Hume. The essential characteristic of the French philosophy of the eighteenth century was opposition to the tyranny and corruption which reigned everywhere —in politics, religion, and private life; an attempt to restore liberty and reason. This spirit of aggression found vent mainly in the person of *Voltaire* (1694–1778), who, though not an atheist but a bitter foe of priestcraft and Christianity, considered a belief in a Supreme Being to be so essential that if there were no God it would be necessary to invent one;

but he doubted the immortality of the soul. *Diderot*, a deist in his earlier days, drifted into pantheism, and regarded immortality as a life in the memory of posterity.

La Mettrie pronounced everything spiritual to be a delusion, and a belief in God equally groundless and profitless. The road to happiness was through universal atheism, which was to put an end to all religious wars. The assumption that the soul lived after the death of the body was to maintain that a function continued after the organ had disappeared. Morality could not be practical unless founded on self-interest, the pursuit of which was man's supreme good. As a physician, he pronounced the brain to have its fibres of thought, just as the limbs had their muscles of motion. Man's superiority over the lower animals was due to the organisation of his brain and his education; otherwise he was a mere animal, and inferior to the lower orders in many respects. Earthly enjoyment was man's morality. There existed nothing but matter and motion, which were inseparably combined; and natural law was eternal and immutable.

This extreme dogmatic materialism, beginning with Locke, closed the philosophy of abstract realism, and naturally created a tendency in the direction of idealism. In realism the tendency is to materialise mind, the spirit being a blank tablet depending for knowledge upon the world of sense—a passive condition; while in idealism the tendency is to spiritualise matter, knowledge being derived from the soul itself—an active condition.

Now reappears the notion of design, pre-established harmony, and this agency is now again invoked to bring about a union between spirit and matter—thinking and being, which characterises the philosophy of *Leibnitz* (1646–1716), a German philosopher and Doctor of Laws, and one of the profoundest thinkers that ever lived. He was master of science and philosophy, and based his physics upon metaphysics. He avoided the dualism of Descartes and the monism of Spinoza through his theory of monads—infinitesimal entities somewhat akin to the atoms in Greek philosophy.

A monad was a simple substance devoid of extension, an indivisible, metaphysical point, possessed of the power of action, and the centre of living activity—a soul, which received

no influx from external sources. Active force was a property
of substance—like a strained bow. There was a plurality of
these monads, the existence of only one being impossible,
and they constituted the essence of all spiritual and physical
phenomena. All monads had ideas which differed only in
degrees of clearness. The thinking monads were capable of
clear and distinct ideas, and could have single adequate ideas.
God was the primitive monad, whose ideas were all adequate,
and, as rational entities, the ideas had consciousness of themselves and of God. In inorganic nature the monads were in
a state of unconscious sleep, only exhibiting motion. The
higher monads, the state of vegetation, enjoyed vitality without consciousness; and in the animal world the monads
attained sensation and memory, the state of reason and reflection being spirit. Between the succession of ideas and the
motion of the monads, there existed through God a predeter-
mined harmony. Man's soul and body stood in harmony like
two watches originally set together, moving precisely at the
same speed. All thoughts were properly innate, being
evolved from the soul, and were independent of external
influences. This world was the best of all possible worlds.

The philosophy of Leibnitz, as well as his theories of
religion and morals, is subject to different interpretations;
but his idealism does not reach its ultimate, and is to some
extent impregnated with realism. We are dependent upon
Berkeley, an Irish bishop, for pure idealism. All our ideas, in
his philosophy, we derived from God, they existed in God,
and there was no such thing as matter. He regarded this
doctrine as the surest means of vanquishing atheism and
scepticism. This dogmatic idealism received no further development; but the philosophy of Leibnitz was further evolved
by Christian Wolff and other German philosophers.

The profound philosophy of the great German illumination
I must but barely mention, for no epitomising consistent with
my limited space can do it justice. It is sufficient, so far as
this inquiry is concerned, to know that these philosophic
problems still remain unsolved; that through abstract theo-
rising we have obtained no knowledge fitted to guide us in
our daily walks; that it has given rise to ever-increasing
discord; and that a different basis for the restoration of harmony is devoutly to be wished. With reference to the political

theories of the leading German philosophers, however, I shall notice them briefly in my chapter on the State.

To *Immanuel Kant* (1724–1804), the profoundest of the German philosophers, is due the credit of restoring philosophy from the unphilosophic extremes of realism and idealism, and through critical inquiry he completely annihilated dogmatism. His philosophy is idealistic, and affirms three postulates, namely, the immortality of the soul, the existence of God, and the freedom of the will. The philosophy of *Fichte* ended in subjective idealism;[1] the philosophy of *Schelling* in objective idealism; and that of *Hegel* in absolute idealism. *Herbert* taught that experience was the basis of philosophy, no knowledge being attainable outside of this limit. With Hegel, in contrast with Schelling, the idea was the absolute, all actuality being merely the realisation of the idea.

The history of philosophy virtually closes here; but in recent years an attempt has been made in England to revive the ancient doctrine of happiness as the supreme good, under the name of the "Greatest Happiness Principle," or Utilitarianism. I would not regard this school as being of much importance in connection with my inquiry were it not for the fact that the tendency of recent political economy, in despair of arriving at unity and truth by the orthodox methods, has been to scatter its scanty elements of fertility over Utilitarian pastures—the basing of economic theories upon another form of abstraction. Another feature is the relation of Utilitarianism to the evolutionary philosophy of Herbert Spencer, who, although in general recognising the Greatest Happiness Principle, is unable to define what happiness is. An essential characteristic of Utilitarianism is, that its defenders are for the most part atheistic in their beliefs, and in this sense it is allied with Socialistic politics and philosophy; while Herbert Spencer, who, although a theist in the sense of entertaining respect, reverence, and awe for an Unknowable, Infinite, and Eternal Energy, is an enemy of revealed religion or scientific theology. His ethics are summed up as follows: "The experiences of utility, organised and consolidated through all past genera-

---

[1] Subjective idealism means that the ideas are produced by the mind; and objective idealism, that they originate in God.

tions of the human. race, have been producing corresponding nervous modifications, which, by continued transmission and accumulation, have become in us certain faculties of moral intuition—certain emotions responding to right and wrong conduct, which have no apparent basis in the individual experiences of utility." This doctrine requires further development before it can be turned to practical account.

The founder of the modern school of *Utilitarianism* is Bentham, who defines utility to be the tendency of actions to produce happiness and prevent misery in relation to a given party, usually the community. Some of his leading disciples are John Austin, John Stuart Mill, and Professor Bain. The postulate of the school is: "The greatest happiness to the greatest number of the human family, with the least injury to any."

Morality, the test of which was utility, was a social instinct, and aimed at the well-being of society in preference to individual happiness. Utility was the only safeguard in legislation. The test of rational morality never varied; the standard was governed by the ability to apply it, modified by differences in education, &c. An action was right or wrong according to its bearing on happiness. There existed a moral sense; but it was not innate or revealed. The useful included the agreeable and the ornamental, but gross evils were inconsistent with happiness. Self-denial, simplicity of living, and tranquillity and excitement in due proportion should be practised. Happiness was increased by removing poverty and disease through the exercise of intelligence and virtue.

Before making any observations on the positive weaknesses of abstract processes, I shall be able to present the question more clearly and forcibly by taking a glance at philosophy's wayward sister, Religion, and shall dispassionately consider her origin and development.

CHAPTER II.

My classification of religion under "abstract methods" may require defence. It will be my duty in this chapter to show that religion, like philosophy, is an abstraction, thus accounting for the various religious denominations ; but I shall be obliged to omit an epitomising of the numerous theological doctrines, partly owing to my lack of space and partly owing to their being well known. I shall not deny that religion is a revelation, but shall also show that philosophy may press the same claims.

A belief prevails that social unity and harmony can never be attained except through religious agencies. It is to be regretted that a subject of such vast importance in relation to my inquiry should be complicated in so many ways. It would be unwarrantable on my part to insist that Christianity is meant when religion is named ; for there are many who do not regard the former as absolute truth, believing that any faith which binds men's consciences serves all the purposes intended, the essential nature of which is a belief in a First Cause. Were I to say that the lack of harmony in religious and theologic thought was proof of a false basis, the answer would be, that there is no want of harmony ; that all civilised nations worshipped the same God, the apparent discord being mere matters of detail; and that the time would soon come when all sects and nations would be brought into a knowledge of the truth.

The following opinions have been advanced with reference to the origin of religions :—1. A feeling of dependence and a conviction of moral obligation. 2. Respect for moral conduct and family life. 3. Fear. 4. Specific ethical feelings, especially love. 5. Desire. 6. The imagination. 7. Poetry (poetic fancy). 8. Wonder. 9. Priestcraft. 10. Quackery. 11. Hero-worship. 12. Ancestor-worship. 13. Use of metaphorical language.

Max Müller, speaking from the standpoint of the origin and development of language, takes the view last expressed. For example : Zeus, the chief God of the Greeks, originally meant the sky; in process of time it became personified, so that in the expression "Zeus rains," the word Zeus came to be conceived as a superior being—a God, who caused the rain to descend, and then other powers gradually became attributed to him.

Although it is certain that our forefathers have worshipped their fellow-men—also stocks and stones, and all kinds of objects, animate and inanimate—yet it may be said, even granting any one or more of the above opinions to be true, that there is no proof of the absence of revelation or inspiration.

The most scientific inquiry that has been made into the origin of religions is by Herbert Spencer, in his work entitled *Ecclesiastical Institutions*, and if he has not presented sufficient evidence to convince prejudiced minds, there is abundant scope for further investigation. The assumption has generally prevailed that religion is a universal instinct, pervading the feelings of all human beings, but this theory has been disproved by scientific investigation. Spencer's conclusion is that all religions originated in ancestor-worship. I shall here briefly summarise the first chapter of his above-named work, giving the main facts without reference to their source. His showings are substantially as follows :—

Man was not a religious being by nature, the theory being disproved both by examination of some uncivilised tribes and also by some civilised people whose minds from infancy had been cut off through bodily defects from intercourse with their fellows. With them neither the idea of a God nor of an immortality was innate. The idea of ghost-propitiation was very strong in the dawn of all religions, and evolved into different forms of worship. In the traces of prehistoric man, all evidence proved that ancestor-worship was the foundation of all religions then prevailing. Amongst some savages, the absence of one's other self in sleep and swoon was regarded as an unlimited absence after death, the spirit commonly lingering near the body or revisiting it, and walking at night in search of food; hence milk was left beside the grave for infants, food and needful implements for adults; and the sacrificing of human beings as food for the dead, sprinkling food about their favourite haunts, &c., became common observances. Unless these "doubles" were attended to in such ways, trouble was apprehended. Some imagined that the dead wandered amongst the living and attended them, and others that the ghosts of their ancestors inhabited the adjacent woods. Food was placed on grave-heaps, and every passer-by added some offering. On a prince's grave (in Vera Paz) a stone altar was erected, where incense was burnt and sacrifices made in memory of the deceased, and shelters for altars, or rude huts, were constructed amongst various peoples, which developed into temples, where praise and prayer were offered. An effigy of a dead man, originally placed on the grave and afterwards in other locations, became a habitation for his ghost; and rude figures of naked men and women became inhabited by spirits. Such images were often propitiated owing to the indwelling doubles of the dead. An African race (Wahebe), however, had no respect for their dead, the bodies being generally cast into the jungle to be devoured by the hyenas. Sometimes the ancestors became identified with certain animals, which were consequently reverenced or worshipped—*e.g.*, house-frequenting snakes—and they were then fed and tended by the descendants of the deceased. Sometimes the ghosts and gods turned into flies. Ancestral names, such as Wolf and Bear, originated in this belief. The stars were camp-fires lit by the dead on their way to the other world; hence was derived the worship of the sun and of the vault of heaven, and names of persons were traceable to other celestial appearances. Deified ancestors grew into superhuman Gods. Emperor-worship developed amongst the Romans, the statues of the emperors being real idols, to whom incense, victims, and prayers were offered. Amongst some of our own ancestors, the Gods were merely superior men. Originally, dream-life was identified with real life, and shadows were entities. Such origins were common to all religious beliefs; and the Hebrew religion was no exception to the rule. The Hebrews placed great faith in the reality of dreams and dream-like beings. The dead sometimes heard and answered, were propitiated by gashing the body and cutting the hair,

received food, spirits haunted burial-places, and demons
entered into men, causing their maladies and sins. Addicted,
like existing savages, to amulets, charms, exorcisms, &c., the
Hebrews also had functionaries who corresponded to the
medicine-men — men having "familiar spirits," "wizards"
(Isa. viii. 19), and others, originally called seers, but afterwards prophets (1 Sam. ix. 9); and Samuel, in calling for
thunder and rain, played the part of a weather-doctor—
a personage still found in various parts of the world. Also
sundry traditions—*e.g.*, the deluge—the Hebrews held in
common with other peoples; and the story of Moses' birth
had its parallel amongst the Assyrians, whose enactments
were of the same general character as those of the Jewish
Code. The Jews believed in and sacrificed to other Gods
than Jahveh, the angels and archangels fighting in heaven,
the people fighting on earth—incidents paralleled by those
of existing savages; and, like other people's Gods, Jahveh
had incapacities, intellectually and morally (1 Sam. iv. 3–10;
Judges i. 19). There were also similarities in the boastful
and revengeful natures (Isa. xlviii. 11); and Jahveh's deception (2 Chron. xviii. 20–22) was likened unto the lying spirit
whom Zeus sent to mislead Agamemnon. So with reference
to the forms of worship, there were also comparisons. Like
the Hebrews, the Mexicans burnt incense; and Jahveh, like
the idol-inhabiting Gods of the negroes, enjoyed the "sweet
savour" of burnt offerings. To the ancient Gods of Mexico
and Central America the blood of sacrificed men and
animals were continually offered up, the image of God sometimes being anointed therewith, and sometimes the cornice
of the doorway of the temple; and the Egyptians and
Greeks, like the Hebrews, offered hecatombs of oxen and
sheep. The offerings of the Greeks and the Peruvians were
without spot and blemish; Jahveh, like Zeus, descended in
the cloud, sometimes with thunder and lightning, and Zeus'
decrees were brought down in tables, as it were. Dancing
was a form of worship performed alike by the Hebrews and
the Greeks, and by various savages. The Greeks regarded
the fulfilment of prophecies as evidence of the truth of
their religion; and amongst the Sandwich Islanders, Captain
Cook's death was a fulfilment of the prophecies of their
priests. The working of miracles was a common occurrence
amongst the Gods of all peoples. Only in the earliest times
did Jahveh appear to men in human shape; the Greek Gods
so appeared but rarely in later traditions, and in ancient
Central America the Gods did not come down and speak to

their devotees, as their ancestors said they used to do. In many religions there were God-begotten sons who descended to the earth; and the Gods themselves came down to ravish our virgins. There was a sort of mediatorship in all religions, and there were many observances corresponding to the Eucharist. The Christian Crusades, to obtain possession of the holy sepulchre, had their prototype in the sacred war of the Greeks to obtain access to Delphi. The Christian worship partly consisted in reciting the doings of the Hebrew God, prophets, and kings; and part of the Greek worship consisted in reciting the deeds of Homeric Gods and heroes. The Greek temples, like Christian cathedrals, were enriched by precious gifts from kings and wealthy men to obtain divine favour and forgiveness; St. Peter's, at Rome, was built by funds contributed in the same way as that by which the temple of Delphi was rebuilt; and, according to Grote, "the lives of the Saints bring us ever back to the simple and ever-operative theology of the Homeric age;"[1] and "various religions in the new and old worlds showed us, in common with Christianity, baptism, confession, canonisation, celibacy, the saying of grace, and other minor observances."

The above abstract leads us to a point of unity in two ways: (1) that all the theories advanced respecting the origin of religion are substantially correct, and (2) that religion is a pure abstraction. It does not reject the idea of revelation properly interpreted, but establishes a kindred quality between religious and philosophic forms of revelation. It would not be amiss to regard both philosophy and religion as revelations, and reject the idea of abstraction altogether.

The shades of difference between philosophy and religion are now readily comprehended. In its earliest conceptions religion is essentially objective, and gradually develops into subjectivity, unless disturbed by potent forces. The dead ancestor or hero becomes an object of the imagination, and as he grows in age and power, the supernatural attributes at the same time increasing, the worshipper grows more and more subject to his will, which is more or less revealed. As the "doubles" correspond on earth, so they must also correspond

---

[1] The words here quoted are not those of Spencer, but those of Grote, the great historian of Greece. The succeeding clause is Spencer's own words.

in heaven; and in proportion to the gain in terrestrial
sovereignty of the tribe, the monotheistic tendency must
harmonise in heaven as on earth. The God who showers
down power becomes a mighty God, and the worshipper
becomes more and more subject to the divine will. This
earthly supremacy, however, enjoyed through ages of triumph,
leads to licentiousness both amongst the worshippers and the
deities, and the latter, still retaining their sovereignty, now
begin to exercise it in the punishment of their devotees on
account of their sins. The philosopher idealises the real
(spiritualises matter), while the religionist realises the ideal
(feels his idealised ancestors). The philosopher shuts his eyes
and thinks (objectivity); the religionist shuts his eyes and
feels (subjectivity). The philosopher's conduct is supposed to
be regulated by actions which he regards as being consistent
with the highest ideas attained by and revealed through his
own reason or that of the school to which he belongs; the
religionist bases his conduct upon obedience to a revealed
will—authority. The question of faith does not separate the
philosopher from the religionist, for he who lacks faith in his
philosophy is no philosopher. All the religious strifes, all the
bloodshed, have been wars of the dead against the living.
The dualism of Church and State, which raged during the
Middle Ages, was subjugated by a victory of the living over
the dead, but the latter are not yet extinct. We have been
governed by Westminster Abbey rather than by Westminster
Hall.

Christianity may now be summed up briefly. The first
striking feature is its lack of originality, although St. Paul
was a great organiser. The licentiousness of the Greeks,
naturally accompanied by a corresponding licentiousness
amongst their Gods, was favourable to a reaction in the direction of the Stoic philosophy, which forms a leading feature
of the Christian religion. The Jewish religion had long been
known to the Greek philosophers, and the commingling of
Jews and Greeks in Alexandria—where Christianity was
born, although cradled in Palestine—was favourable to an
amalgamation, especially as the most highly cultivated
amongst the Alexandrian Jews had now continued the

practice of interpreting the Sacred Scriptures allegorically. The policy of the Jewish-Greek philosophers was to blend Judaism with Hellenism; and Philo, who was the first to reduce theosophy to a system, allegorically introduced philosophic ideas into the Jewish religion. Scepticism, which developed into Neo-Platonism, formed a strong element in Christianity. Christ was an attempted union of realisation and idealisation. The Incarnation typified Jewish (subjective) realisation, and the Ascension typified Greek (objective) idealisation. The social forces produced an irresistible tendency in this direction. There is nothing surprising in this; for, as we shall see, the abstract thinkers of the present day are making desperate efforts to force natural laws into lines parallel with their ideas. Such violent dualisms possess no element of perpetuity. The early success of the Christian religion was largely due to the favourable impression which it made on the poor, who, through the promises of Heaven, heard the Word gladly, and partly to increased facilities for spreading knowledge. Comfort is always welcome in times of poverty and oppression. The Church, taking advantage of the new situation, preyed upon the superstition of the masses in order to subject their consciences, which ended in ecclesiastical suicide and religious individualism, just as feudalism ended in political individualism, and the Socialistic reaction of the present day must prove the beginning of a similar cycle. The poor of our times are reversely situated, it being now the policy of the rich to hear the Word gladly, and the history of these cycles of objectivity and subjectivity prove the futility of attempting to maintain order and harmony through abstract methods. Christianity may be characterised as a commingling of Judaism, Jewish-Alexandrian philosophy, scepticism, Stoic materialism and self-denial, Neo-Platonism, and scholasticism. Such an admixture of monisms and dualisms is doomed to a tragic fate. In our own day we can distinguish the predominating elements. For example, in Romanism and Anglicanism we find sacerdotal Christianity—Jahvism; in Presbyterianism, philosophic Christianity—Jovism; and in Methodism, ecstatic Christianity—Neo-Platonism. All our religious denominations are varying shades of these three forms.

Many curiosities may be presented in another shape. For instance: the Stoics, materialists, founded the most spiritual philosophy and led the most spiritual lives; Locke, a Churchman all his days, was the founder of modern materialism; Spinoza, a Jew who rejected Judaism without embracing Christianity, also a materialist, led a profoundly pure, pious, and abstemious life; the Christian Church, spiritualistic, sank into the lowest depths of corruption; and Socrates, a heathen philosopher, was inspired to preach the Gospel, disdaining any reward except a pure conscience, and chose death rather than a life inconsistent with his religion.

Let us now consider a few of the weaknesses of philosophy and religion from the positive standpoint. First, and not least, are the arguments with reference to design and a First Cause. In concrete methods of thinking there can be no such thing as a First Cause—that is, a first cause of all things. Our little world, even our solar system and many other systems, may have a first cause; but neither matter nor force (including spirit or mind), or any combination of them, can beget itself, even by virtue of its own innate, supernatural intelligence. In examining concrete objects, we find a Last Effect to be just as imperative as a First Cause—that every cause must have an antecedent, as well as a subsequent, effect. Simple addition and subtraction teach us this truth. We cannot add one marble to another without subtracting it from something else, nor can we add a shovelful of coal to the flame without subtracting it from posterity. Equally fallacious is the argument of design. I have already pointed out instances, and shall point out many more, in which nature shakes off dualisms and restores harmony, law, and order in spite of temporary resistance from civilised man. Harmony, or marks of design, is a condition of existence. A watch is a more ingenious contrivance than a rat-trap, and the question resolves itself into this: Which is the easier belief, that the watch, or the trap, made itself? An answer to this question will enable us to decide which is the more rational belief, that the world or the supposed First Cause is self-existent. Besides, it is impossible to conceive an era in which neither time nor space existed, and if we could conceive the existence of nothing but space, we could not think of it as being an absolute vacuum, containing neither heat nor cold, matter nor force, light nor darkness, or as being bounded by a wall. Another illustration will show how easily we can infatuate ourselves by fanatical adherence to abstract theories. The ancient Greeks believed that their giant-god, Atlas, supported the world on his shoulders. Who or what supported Atlas? There is a tradition, probably modified from an ancient Indian source, that the world rested on an elephant. What does the elephant stand on? On another elephant. What does the other elephant stand on? Oh, there are elephants all the way down. Here are three different periods in the development of this religion, but the tradition leaves us, I suppose, to infer that the last elephant stood on the back of one so large and strong that he was able to support all the smaller elephants and the world on their backs. This religion was more highly developed than that of the Greeks, who could only discover one elephant (Atlas); and the Christian religion is more highly developed than the Greekish, as the following facts will show. Philo, a Jewish-Alexandrian philosopher, regarded the Word (*Logos*) as being intermediate between God and the world, God being the Cause of the world as distinguished from the Word through whom it was formed. Marcian declared the Creator of the world to be inferior to the supreme God; and Irenæus taught that the *Nous* (personified reason) emanated from the unbegotten Father, the Word emanating from the *Nous*. The early Christians taught that the world was created or formed by the Word whom God created for that purpose. In these examples we find our world standing on two elephants.

There is a weakness which pertains more to religion[1] than to philosophy. I refer to the allegorical interpretation of the Scriptures. This weakness, however, is also a source of their strength. If they were incapable of such interpretation, the march of thought, freed from the fetters of religion, would utterly destroy them in a scientific age; but so long as they can, by flexibility of interpretation, be made to follow the

---

[1] As will be seen in a later chapter, the religion of Christ has a true scientific basis, but priestcraft "religion" is quite a different matter.

progress of events, their existence is more secure. By its very nature, religion must be strongly conservative, and can only be moved by a strong and persistent force, while progressiveness is not so inconsistent with the nature of philosophy, despite our stubborn adherence to the "wisdom of our fathers."

When scholastic philosophy, called to aid religion, failed, it was urged by the early reformers that no Christian could philosophise correctly unless he received spiritual regeneration. This anticipation not being realised, the Church received another blow; for one of two positions was proved: either (1) that the rule was unphilosophic and false, or (2) that the priests and other expounders of Christianity were not spiritually regenerated—because the breach grew wider and wider after this declaration.

Rev. Dawson Burns says: "The whole machinery of religious instruction, were it multiplied a hundredfold, would be insufficient in acting upon the mass of misery and immorality which is drink-created." Professor Blackie, who is certainly orthodox, praises Socrates for his piety,[1] the inference being that piety from all sources, if producing equal intensity, has equal value—that heathen piety, if it produces the same results as Christian piety, conduces to the same end. I recently read in a monthly magazine a remark made by a clergyman (the names have escaped my memory) to the effect that barbarians perished through forces from without, and civilised society from inner decay. This is a sad commentary on Christians, who maintain that civilisation is the offspring of Christianity, and we arrive at the absurdity that the same force which is intended to enquicken our inner life is productive of its inner decay. I make the observations contained in this paragraph for the purpose of showing the weakness which results when the preachers of Christianity lose faith in the efficacy of their own treatment.

It is certainly a practical weakness when the various religious sects cannot be classified according to their moral standards. When a position of trust is vacant, who is most competent to fill it, the sacerdotal, the philosophical, or the

<hr>

[1] Four Phases of Morals.

ecstatic Christian ? The futility of attempting such classifications, together with the fact that the employer does not concern himself about the theological beliefs of his responsible *employé*, greatly detracts from the practical efficiency of religion. From the practical standpoint, Christianity is subjected to another weakness. Statistics show that over £1,000,000 are raised annually in this country for Protestant missions, and it is estimated that an equal sum is annually raised in Europe and America. About 6000 European and American missionaries and 30,000 native agents are employed in Christianising the heathen. The annual increase of native Christians, due to missionary zeal and enterprise, is reckoned to be 60,000; but we have no estimate of the sums required to prevent relapses and backslidings. It may be tolerably easy to persuade the heathen that the shades of the Jews' ancestors are more powerful and dreadful than the ghosts of his own ancestors; but to maintain him in this belief demands piety, skill, and money. This circumstance would not be so misfortunate for the cause of religion were it not for the fact that these £2,000,000 would maintain annually the bodies and souls of over 200,000 of the starving heathens in our own cities; and, besides, no sums have been collected for the conversion of our millions of highly educated barbarians. Money collected for missionary work should, one would think, be no respecter of persons.

Abstract theories are also subjected to prospective weaknesses. Granted that the philosophic dualism is solved, will abstraction cease here ? Be it conceded that man, both body and soul, is a material substance, then a new set of philosophers will spring up with the proof that he is the deadest of all material substances, there being no mind or motion in him. Be it proved, on the other hand, that man is all spirit, there being no such a thing as matter, then sophists will arise with infallible proofs that he is merely the shadow of a spirit —that ideas are the ghosts of men's souls. Granted that a conciliation occurs, the soul living in perfect harmony with the body, then the logical conclusion must certainly follow, namely, that body and soul are so intimately related that no force, human or divine, can rend them asunder; that death is therefore a pure fiction ; that pain, poverty, disease, &c., are idols of our diseased imagination, so that the poor should be contented with their lot wherever it may be cast. Granted, again, that philosophers find the object of their chase, namely, the First Cause, they will then begin to wonder what this elephantine Atlas stands on to ease the soles of his weary feet. Granted even once more that our philosophers are a unit in the belief that happiness, or any other abstract quality, is the chief end of man, then real discord first begins. What is happiness ?—a question on which no two individuals can agree, not to mention the various bemuddled schools of philosophy. All this goes to show that there is no end to abstract processes, and that through them no practical results can ever be attained—nothing to make us wiser or better men.

I shall not, however, dwell upon the positive weaknesses of abstract methods, believing that I have a good case in the negative side of the question—that after thousands of years of profound investigation and sanguinary conflicts opinions are growing more divergent, and that the times are ripe for the adoption of another basis, there being only one other, namely, the scientific, concrete, or mathematical. It is now my duty to preserve religion and philosophy from utter ruin by solving the dualism of subjectivity and objectivity.

Referring to my remarks on axioms, it is here in place to establish the first of my series, viz. :

Axiom I.—*No structure, no function ; no function, no structure.*

Deductions.—If our race perish structurally, the human soul perishes also. If our race had never existed structurally, the human soul would never have existed.

I confine the functions to the soul because, although Christians have denied that there is any essential connection between the soul and the body, it has never been·denied that our other functions—*e.g.*, wisdom, happiness, strength, virtue, vitality, motion, &c.—can exist without structural man. If it be asserted that any one or more of the above-enumerated qualities are functions of the soul, then we are fixed in one of two positions : (1) that the soul is a material substance, or (2) that such a condition exists as the function of a function. It is not necessary for me to assert or deny any one of these positions ; my duty is to adhere to method. My position is this, that the theory of the structure possessing anything beyond a function has proved a failure, and that such a condition, if it exists, is beyond the range of science. The assumption of this condition, if varying results in scientific inquiry be thereby produced, would put an end to all scientific investigation. The above axiom must be criticised from the scientific, not from the abstract, standpoint.

If a man suddenly lose his life without violence to his body, the structures are left, but the functions are gone. When the body is resolved into dust, the structure disappears as such, although no matter or force becomes annihilated. Again, a living structure may change its function, which, if the change be not too violent, causes also in process of time a corresponding structural change ; but if the structure ceases to exercise its natural function, or assume some new function, it is only a matter of time when the structure will disappear. The same law applies to an inefficient exercise of function, by which the structure becomes atrophied, thus deteriorating in size, weight, and strength ; but in this case a greater period of time is demanded for the disappearance of the structure. The discontinuance of any vital function ends in the destruction of all the bodily structures and functions. By exercising the structure to its fullest capacity without strain, the maximum degree of development is attained, providing the exact quantity and quality of nutriment necessary for such a purpose be supplied, and the structure then gains in size, weight, health, and strength.

CHAPTER III.

POLITICAL ECONOMY.

In the previous chapters we have been discussing abstractions
and revelations, and those who still have faith in such processes will do well to follow me in this chapter of abstractions and negations, in the hope that some redeeming feature
may be realised.

Although, in its highly-developed form, economics is a
young theory compared with religion and philosophy, and has
not yet had time to unfold so many fallacies and divergent
schools of thought, yet I feel confident that a brief survey of
the origin and development of political economy will enable
us to place reliance on the negative side of the argument, as
I have done with philosophy, although I shall also point out
many positive weaknesses. By this method, namely, showing
the absence of unity and harmony amongst our economists,
I hope to be able to dispense with apologies for shifting the
basis from the abstract to the concrete.

In the solution of social problems the mind naturally tends
to gravitate towards economics, wealth being the mainspring
of our actions, and the power in and behind our throne.
Economicism holds within its iron grasp both the Church and
the State, and has absorbed science and philosophy.

Going back to the most ancient history, we have little to
learn in economics from the Oriental Theocracies, and to draw
a sharp line between the economics and theology of that
period would be a laborious task. For the first principles of
economics, as well as of philosophy, we must return to ancient
Greece, and even there, where politics, economics, ethics, and
theology were all united in the State, a clear separation is
practically impossible. What we do know is that, in ancient

as well as in modern times, there have been slaves and free-men—call them Plebeians and Patricians, Roundheads and Cavaliers, Whigs and Tories, if you like—and there have been in all ages economic disturbances, class privileges and abuses, riches and poverty; priests and landlords plundering the poor and then returning a portion of the booty in the shape of alms; oppressive taxation, fiscal and military annoy-ances, and all the other great incidents which combine in developed form to build up our modern civilisations. It does not alter the principle to say that ancient or mediæval slavery was of the militant or chattel kind, while modern slavery is economic.

The principles of political economy were well known to the ancient Greeks, but not in a systematic form; and they became so entangled with the political institutions that the historian cannot easily locate the abuses of those times. Pro-perty, not being safe under a spirit of military aspirations, did not tend to accumulate and develop into capitalistic sove-reignty, as in our day; and it belonged, as a rule, to the State. The public exchequer was the source of revenue for every individual or family in the State, and in the main this fund was maintained by slave labour. The moral impulse, as well as the " filthy lucre," emanated from the same external source —the State ; but the coin was theoretically to be devoted to the high and moral aims of the public, and not to be esteemed for the sake of personal enjoyment. There was an Athenian doctrine that no citizen should want; but this political theory, carried into practice, was productive of idle and improvident citizens. Foreign trade was discountenanced and discouraged, although the Athenians derived revenues from custom duties as well as from mines and public domains. Like modern nations, the Greeks delighted in taxing foreigners; and although they had their tithes, they had no door or window tax, and had a strong antipathy against land and personal taxes. Smuggling was quite a common practice, and manufacture was discouraged. Xenophon held that most of the manual arts tended to deform the body; and Plato, himself a great athlete in his youth, was as clear as Adam Smith on the advantages and disadvantages of the division

of labour in manufacturing industrialism. Aristotle, in harmony with other Greek thinkers, was a physiocrat, believing
that in agriculture and kindred arts reposed the real dignity of labour, and these were the truly productive employments. Xenophon, in direct opposition to the adherents of
our modern Mercantile System, did not believe that the export of money, which mainly consisted of gold and silver, in
exchange for commodities, would impoverish the community ;
and Plato forbade the obtaining of interest for the use of
money, even the repayment of the principal being left to
the option of the borrower. There were laws against usury,
monopoly, and money speculators. The Greeks were strict
Malthusians, not from fear of the failure of the means of
subsistence, but on moral and political grounds ; and for the
purpose of restricting population, early marriages were pro-
hibited, means also being taken for the prevention or destruction of births. It is not reasonable to suppose that economic
theories could enjoy a high state of development amongst a
people who condemned economicism.

Amongst the Romans there was a dearth of economists, as
well as of philosophers, their time being fully occupied in the
development of their military and political institutions, but
they paid considerable attention to agriculture, which was
chiefly prosecuted on large estates by slaves acquired through
conquest. Their social and economic ideas were essentially
Grecian. Industry and commerce were regarded as degrading
occupations, unworthy of free citizens. Pliny preferred barter
to the use of gold as a medium of exchange ; and both he and
Cicero advocated the prevention of the exportation of money.
Interest was fixed by law, and money-lending was regarded as
an odious occupation.

The mission of the civilised world during the Middle Ages
was the elaboration of ecclesiasticism and feudalism, there
being no room for the development of manufacture and trade,
and even agriculture lagged behind. As amongst the ancient
Greeks, a large portion of the wealth was squandered in the
maintenance of ecclesiastical ceremonies ; and a large percentage of the remainder was wasted in supporting a large
body of retainers under feudal customs. At a later period,

however, the Crusades gave a stimulus to international trade. The Church preached the virtues of industry, thrift, obedience, the dignity of labour, and almsgiving, but discouraged trade, because it led to the perpetration of frauds. Attempts were made to fix the prices of commodities by law, interest was forbidden, and slavery was transformed into serfdom. The serf, though tied to the soil, enjoyed certain religious and domestic privileges. Money played an insignificant *rôle*, and wages were mainly paid in produce. The goods were mostly made in the dwelling-houses for consumption by the family, not for sale. People usually consumed only such products as were raised on their own land and prepared in their own houses. Tradesmen either worked in the houses of the consumers, or made goods to order in their own houses, and were usually paid in produce. Indeed, the officials, literary men, artists, poets, &c., who served the great landlords, enjoyed board, residence, and clothing from their patrons, or received a piece of land, with free help, from which they obtained their subsistence. Even the taxes were paid in produce and services. The farmer, for example, gave a portion of his produce to his landlord, and worked at occasional intervals for him on the estate, or was employed in making roads. The landlord, instead of paying taxes directly to his superior lord, the king, rendered military services when necessary without compensation. During these times there were no great commanders or standing armies, and the authority of the superior lords was comparatively limited. The great landlords inherited the offices of judge, military chief, governor, chief of police, tax-collector, &c., within their respective circuits, comprising a greater or less number of towns and villages, but received no reward for their services, and it was very difficult, or even impossible, for the superior lords to remove them. The latter, again, were subordinate to kings. There were trade guilds, who regulated all the affairs between employers and employed. They decided the number of apprentices, the methods of working, kinds of tools, inspectorships, length of working day, &c.; they partly made joint-purchases of the raw materials, and joint-sales of the finished products, conducted insurances, dispensed charities, &c. They built a sort of State within the State, and made their own laws, the laws of the State being very restricted, and carried out to the letter the motto, "Each for all, and all for each."

Such, then, briefly expressed, was the condition of our ancestors previous to the breaking out of modern civilisation, ushered in by the sovereignty of the Mercantile System, the rule of money, the marvellous development of economic theories, and all the other blessings which have crowned the nineteenth century with the precious jewels of "liberty, equality, and fraternity."

The strict adherents of the school of political economy known as the *Mercantile System*, which attained the height of its development about the middle of the seventeenth century, taught that money and wealth were identical.

Money was always in universal demand, and gave the possessor power to purchase all other commodities. It was therefore the duty of the State to resort to all possible expedients—prohibitions, duties, bounties, restrictions, &c.—for the purpose of retaining in the country as much of the precious metals as possible, thus bringing about a favourable balance of trade. The more moderate mercantilists, however, whatever their private opinions might have been, did not express themselves so radically with reference to the identity of money and wealth. The Mercantile school assumed the responsibility of teaching the governments what course they should pursue in order to make the nation rich in the means of subsistence, and consequently also populous. A country which pursued agriculture alone could not, without manufacturing industries, develop the resources of the country and the faculties of its inhabitants, and it must therefore remain poor and sparsely populated. It was, therefore, the duty of the State to open up industries and commerce, and furnish the inhabitants with employment, thus procuring for themselves the means of subsistence, instead of depending upon other countries through the importation of foreign productions. The importation of foreign commodities was to be prevented, or rendered as difficult and precarious as possible, by heavy import duties, while exportations were to be encouraged by freedom from such restrictions and the granting of bounties—except raw materials, the exportation of which was to be limited by the imposition of export duties, the importation thereof being facilitated by exemption from taxation, thereby causing the home-manufactured articles to be cheap and the foreign articles dear. The same policy was to be pursued with reference to grain, by which the labourers could live and work more cheaply, thus producing cheaper articles for export. Special concern was to be exercised by the State that the balance of trade was favourable, namely, that more money was imported than exported. The export of gold and silver, coined or uncoined, was to be rendered as difficult as possible; the import, however, to be correspondingly encouraged. There was to be no economic wall between different parts of the same country.

There are three features in this economic school on which I desire to lay special stress, being of great importance in connection with this inquiry, namely: (1) the politicians of the sixteenth and seventeenth centuries undertook to enrich the nations in the manner prescribed by the economic theorists; (2) the encouragment given to population to enjoy the immense wealth to be produced; and (3) the discouragement of agricultural pursuits and agricultural knowledge.

The immediate result of Mercantilism was the abolition of the feudal system of economy, the conversion of agricultural into manufacturing and commercial communities, and a strengthening and centralising of the governments. Powerful and populous cities sprang into existence, demands for men and money for the maintenance of standing armies and religious wars were increased, Court extravagance, political patronage, and public officials attained an extraordinary extreme, and, owing to the dwindling of the royal domains, taxation grew very oppressive. Politicians saw that this system tended to increase their powers, and they gladly undertook the control of industries, or granted monopolies to privileged corporations, and the trade of colonies was restricted to the mother-country. The landlords, losing interest in their avocation, were drawn to the great centres of luxury and pleasure, where the profits of agriculture were squandered; and the rulers, driven to extremes of pride and prodigality, resorted to debasement of the currency to maintain their royal dignity at the expense of the toiling masses. Naturally enough, these theories were formulated into a science (!) by the. literary and philosophic geniuses of the

day. Prominent amongst the European rulers who adopted
the Mercantile System may be mentioned Cromwell, Charles
V., and the French Minister Colbert. In a milder form,
Mercantilism rules the civilised world to-day, and is not yet
extinct in Britain.

Amongst the early opponents of the Mercantile System
may be mentioned Sir Dudley North,[1] who taught that wealth
might exist independent of gold or silver, its true source
being human industry either in agriculture or manufacture.
He acknowledged, however, that the precious metals played
an important part in the wealth of a nation. Trade stagnation, he taught, arose not from a lack of money, but from
a glutted market, a disturbance in foreign commerce, or a
restricted consumption occasioned by poverty. The export
of money tended to increase, not to diminish, the national
wealth. No nation could become rich by State regulations ;
peace, industry, and freedom were the sources of prosperity.

Locke, although stoutly resisting the debasement of the
currency, was strongly imbued with the virtues of Mercantilism. He regarded a fall in land rent as a sure indication
of a decline in national wealth.
Like the philosophic and religious dualism discussed in
the previous chapters, we now find, as an identical law, an
economic dualism. When one extreme is about to be reached,
there can only be suicide or reaction. The reactionary
phase of the Mercantile System is *Physiocratism*, the Physiocrats deriving their name from two Greek words meaning
natural law, which they regarded as the basis of their system
—like the motto of the Stoics : Live harmoniously with
nature. This new French school of economics adopted the
motto : *Laissez faire*—the *Geschehen-und Gewährenlassen* of the
Germans—which is faintly expressed by our phrase, State
non-interference ; or, as we put it, State interference. It
means, when applied to economics, that the State should not
interfere with the industrial or domestic affairs of the nation,
that trade should be left to flow in its natural channels. This
dualism of Mercantilism and Physiocratism is identified with
that of Socialism and Individualism of the present day, the

---

[1] Discourses upon Trade, 1691.

curiosity of which is that the individualistic aristocracy of to-day were the Socialists a century and a half ago. The Physiocrats proper only taught the *Laissez faire* doctrine at first; but there was another school known as the Agrarians, who espoused the cause of agriculture, the two schools having been united by Quisnay (1694–1774), physician to Louis XV., who spent his early days in agricultural pursuits. Gournay (1712–1759), the most distinguished founder of the Physiocratic school, was a merchant and a mercantile philosopher. The economic development just prior to this period consisted mainly in the division of banking and commerce into separate branches, and the extended use of machinery. Pierre Bois-guillebert taught repeatedly and with great emphasis that national wealth did not consist in gold and silver, but in utilities for consumption, the greatest of all being agricultural products, and that State interference did more harm than good. Marshal Vauben, appalled at the deplorable condition of the labouring classes, taught that it was the duty of the State to promote the welfare of all classes of the community; that labour was the basis of all wealth, of which agricultural labour was the most important; and that all unprovoked restrictions on manufacture and commerce should be removed, protesting at the same time against the privileges and exemptions enjoyed by people of high social standing. This was the condition of affairs in France when the Physiocrats systematised their doctrines, which, briefly expressed, are as follows :—

The State had no moral right to do more than protect its subjects from attack, and enable them to enjoy the fruits of their own labour and that of their fathers. Agriculture was the only productive industry. Although trade and the other industries were not productive, yet they were useful so long as they served agriculture, and made no higher profits than enabled them to do greater services to agriculture without enriching themselves. Only those occupations were productive through which those means were obtained from which men could live, and then only so far as such occupations produced more than was required to enable the workers to live and continue their operations. The extent of the world's population depended upon the quantity of food, fuel, clothing,

&c., produced from year to year. How many others besides the tillers of the soil could live depended upon the quantity of the necessaries of life produced by the farmers after supporting themselves and making provision for the continuance of their operations. This surplus determined the number of landlords, merchants, tradesmen, manufacturers, professional men, &c., who could live. By free-trade, under the influence of free competition, commodities commanded their natnral prices, namely, such prices as covered the cost of production, enabled the producers to continue their operations, and educate their families for the same employments as their parents. Labourers would then also not receive more or less wages—the natural wages—than enabled them to obtain the means of subsistence, and educate their families for the same class of work. Merchants would also, in consequence of free competition, only receive sufficient profit to enable them to continue their business and bring up their children in mercantile pursuits. Only the landlords enjoyed special privileges, and were enabled to live in luxury and ease, because it was not possible for them to have free competition, so that they were not reduced to the bare means of subsistence. They should therefore pay all the taxes, and fill all the higher offices of State, civil and military, without compensation. In a naturally regulated State there should only be one form of taxation, namely, a direct levy on ground rent.

The Physiocrats made little impression in Britain, have not been highly favoured in Germany, have never attained much influence even in France, where they originated, and their doctrines have played no part in the creation of the existing condition of our social affairs, either on the Eastern or the Western hemisphere. This fact may be attributed to the poverty, ignorance, and depression of the agricultural masses, who have therefore been unable to study their own interests; and their calling, having been pursued by slaves and serfs, was brought into disrepute.

The dualism is now complete, namely, economicism and agriculture. It is a struggle for the supremacy of luxuries on the one hand, and of the necessities and necessaries of life on the other. The economist makes manufacture and trade the basis of national prosperity, and the physiocrat takes agriculture as the basis.

Before passing on to the next economic era, a word is due to Hume, as it has been to Locke, in order to show that economic theories have been moulded by philosophic thought, which is the direct reverse of scientific methods of thinking. Hume, although repudiating the Mercantile school so far as it confounded money with wealth, was generally an adherent of Mercantilism, and he was a moderate advocate of protection to home industries, although condemning extremes in this direction. He repudiated the single tax on ground rent as advocated by the Physiocrats.

Another distinguished predecessor of Adam Smith, Dean Tucker, although supporting Free-trade doctrines, favoured bounties on exported goods, and upheld the encouragement of population by a tax on celibacy.

The next revolution in economicism was the *Free-trade* theories of Adam Smith (1723–1790), a Scotch philosopher and theologian, who, like the ancient Greeks, mixed up theology, ethics, politics, and economics. His theories may be summed up as follows :—

The idea of wealth was to be found more in industry— diligence in agriculture, manufacture, and commerce—than in the soil and in nature. Those occupations were productive which gave the raw material of nature an exchange value and a market price. But those occupations which only rendered personal or intellectual service—those of teachers, lawyers, doctors, judges, &c.—were not productive in the economic sense. The increase of national prosperity consisted essentially in the growth of diligence, dexterity, and skill. The extent to which the division of labour could be carried depended on the extent and purchasing power of the market. The trading instinct of man led to the division of labour. The main conditions in the development of industry were security of person and property, perfect industrial and commercial freedom—both internal trade and foreign commerce— and the equipment of labour with capital—*i.e.*, supplies of food, raw material, &c., and especially tools and machinery. Capital could only originate through saving—the withholding from immediate consumption a portion of the products and employing them in further productive operations. Men rendered greater services to their fellows through the system of bargain and sale than through philanthropic motives. Self-interest was the surest incentive to diligence. The greater the competition, the greater the incentive and necessity to undersell competitors in order to retain custom. Utility alone, or value in use, did not decide what quantity of one article was exchangeable for another: those articles which essentially had exchange value were the products of labour. So long as landlords existed, as well as labourers, the labourer could not get the full produce, or the full exchange value, of his labour. Without working himself, the landlord received a portion of the products of labour for permission to use the land, even when it remained in its original, uncultivated condition. The price of the products must then, in addition to the wages, yield a ground rent, before production could be carried on. When people accumulated capital from their sparings, wherewith they employed labourers, then a portion of the total produce, or its price, fell to the capitalist as capital profit. This was divided into two parts: (1) the actual profit for the man who invested the capital at his own risk, and (2) the interest for the loaner of the capital. By the law of free competition, there existed a mean condition to which the prices of the goods, as well as the profits, converged. When the prices or profits were high, fresh capital and labour rushed in until the surplus profits disappeared. On the other hand, when prices and profits were below the level, capital and labour forsook the occupation. The extent of population was largely determined by the demand for labourers. The number of the labour population determined the death-rate amongst that class, particularly of children. When wages for a considerable length of time (fifteen to twenty years) remained under the margin of subsistence, the numbers sank, largely through the death-rate amongst children, the results of bad and insufficient nourishment—also through reduction of families grounded. For a long period the wages could not, therefore, remain below the margin which enabled the labourers to exist, and raise children for working purposes. High wages were desirable, for the labourers numbered the largest portion of the population, and good food and care were promotive of working efficiency; so that, also, excessive diligence, as by job-work, was injurious to the labouring classes, reducing their numbers. In sympathy with the prosperity of the nation, both wages and rent rose, while profit and interest sank. In periods of national adversity wages and rent sank, while profit and interest rose. The interests of the labourers and the landlords therefore harmonised with the interests of the nation; for all conditions which increased the prosperity of a nation enhanced rent and wages, and all the conditions which opposed the nation lowered rent and wages. On the other hand, the interests of the capitalists were opposed to those of the nation, for profit and interest sank when prosperity increased, and rose in times of national adversity. All labourers' unions were detrimental to the interests of the State, even though they were only intended to support impoverished members and like objects, because they acted injuriously to the interests of the consumers, and led to an increased price of the products of labour. It was the duty of the State to raise by import duties such sums as were required to maintain its military strength, and to encourage educational institutions, ship-building, fisheries, &c.

It will be observed that Adam Smith's school of political economy is physiocratic, but not agrarian; and, by a cursory inspection, it may be inferred that Smithism is an attempt to conciliate the dualism of economicism and agricolism—the agricolo-economic dualism. Such inference, however, is a misconception. Smith bases national prosperity on manufacture and trade, and his belief is that economicism can be better promoted through Free-trade than through Mercantilism. Free-trade is the subjective phase of economicism, Mercantilism the objective phase. Previous to Adam Smith, the old Physiocrats taught that God had so adjusted nature that, if her laws were left undisturbed, all men could live on the earth in a state of concord and prosperity.

The next great magnet in the sphere of economics is the Rev. T. R. Malthus (1766–1834), the reputed father of *Malthusianism*. This distinguished theologian also, as a disciple of the school of discord, defended the landlords against the reproach that they enriched themselves at the expense of the labouring classes, and were the cause of so much misery in the world. His theories may be briefly summed up thus:—

The greater the population the greater was the difficulty in procuring the means of subsistence. The prices of grain must then rise, and those portions of land which were specially productive must also rise in value and bring more rent. However, while it became more difficult to produce food, luxuries and other products of industry were more easily increased. It was not, therefore, the high rents of the landlords and the luxuries of the rich that were the cause of the misery of the masses. The source of the evil lay in over-population, or in the laws of nature which caused the means of subsistence to lag behind the requirements of the people—that "population increased in a geometrical, food in an arithmetical, ratio;" so that it was the duty of everybody to become conscious of the fact, and fulfil the duty of ceasing to found a family till he had acquired a sufficiency for their support, or had prospects sufficiently bright to maintain his family. This check to over-population was to be brought about by "moral restraint"—abstinence from marriage till late in life. He who brought children into the world without ample provision for their subsistence should be left to stern retribution, and, owing to such a miserable ambition to defeat the laws of God, both he and his family should be doomed to suffer the terrible consequences.

At the present day we have a theory under the name of *Neo-Malthusianism*, which was not dreamt of by Malthus, and is a natural outcome of the development of luxury. Marriage, like all other luxuries, must now be indulged in by all classes, and the idea of preventing conception after marriage is now entertained as a remedy for over-population—increasing the number of early marriages and limiting the number of births by artificial arrangements. Mill's theory was post-nuptial continence, which was also remote from the mind of Malthus. Another school of modern economists, notably the Socialists, hold that food tends to increase faster than population, that man's power over nature increases faster than any possible increase of population. The distinguished economist, Senior, was also an anti-Malthusian. These theorists blame monopoly for the existence of apparent over-population.

The essential character of the next shift in political economy is the *Wage-fund* theory, associated with concord by David Ricardo (1772–1823). In this economist we find the practical business man, the speculator in stocks, therefore not the philosopher or theologian, although his mind was biassed by the abstract methods of his predecessors.

He agreed with Malthus with reference to the cause of misery amongst the labouring classes and the source of rent. Although his theories harmonised in the main with those of Adam Smith, especially elaborating the theory that labour was the source of value, yet he attempted to show that the landlords became enriched without rendering suitable services in return, and that the interests of the capitalists and those of the working-classes harmonised. By the operations of nature's laws, wages, in the long-run, could not be higher than would enable the labourers to perpetuate their class without increasing the numbers. The misery of this class could only be eradicated when capital in general (outfit of raw material, tools, machinery, &c.) and the wage-fund (the means of subsistence for the labourer) could be increased more rapidly than the labouring classes, or when these classes proportionately increased their numbers. It was the capitalists who created the wage-fund by their sparings, from which the labourers were enabled to subsist. The more the population increased, the greater was the tendency to bring less fruitful soils under cultivation. The cost required to produce grain under the most unfavourable conditions at a given time determined the price thereof. In more favoured localities a surplus over and above the cost of production was returned, which fell to the landlord as rent, even when he did not work himself and only permitted the tenant to use the land. The greater the increase of population the higher the price of grain, and the higher the rent of the landlords. The State should therefore suppress those influences which tended to raise rent, such as import duties on grain, and should embarrass rent as much as possible, whereby, in the interests of the working-classes, the capitalists would be free from burdens; for the more they were relieved from burdens, the greater grew the capital, and consequently the wage-fund also.

Amongst the opponents of the wage-fund theory may be mentioned Professor Jevons, who regards it as "illusory"—except that "it may have a certain limited and truthful application;" and Henry George, who repudiates it peremptorily and without compunction, contending that the labourers are supported from the produce of their labour, and not from capital. Amongst the earlier economists who opposed the wage-fund theory may be mentioned Cliffe Leslie and F. D. Longe.

It will have been observed that the Smithian and Malthusian theories favoured the landlords, while the Ricardian favoured the capitalists, and the trinity were a unit on the theories of population and free-trade. Ricardo's system of economics was expounded and elaborated by John Stuart Mill, the distinguished logician and Utilitarian philosopher, and also by Professor Cairnes, a disciple of Mill. This virtually closes economic orthodoxy in England, although my inquiry impels me to notice a new school, which I shall do later on.

The capitalists found other champions of their cause in America and Continental Europe.

This school, known as *Protectionists*, taught that it was the duty of the State—and it therefore exercised the right—to foster home industry by a tariff imposed on foreign goods entering the country, thus lessening the competition of home producers, who suffered through importation of the cheaper goods from abroad. Such a policy would have the effect of embarrassing importations and of giving employment and food to many people in their own country, instead of depending upon articles from abroad, which could, through encouragement, soon be produced more cheaply at home. The free-trade theorem, namely, "Buy where cheap, and sell where dear," did not always work; and the manufacturers should receive encouragement until their business was so far developed that they could compete with foreign countries without protection. Free-trade was only justifiable so long as the country enjoyed a strong exporting power, which enabled it to conduct its commercial operations without being endangered by foreign competition.

The great champion of the Protectionist school in America was Henry C. Carey (1793-1879), a publisher, who taught the following theories :—

It was a law of nature that the ability of a given species of animals to increase its numbers was so much the greater the lower it was in the scale of existence. On the contrary, the higher the species in the scale of perfection, the less rapidly it multiplied. The lower the animal existence, the more it served as food for the higher animals; so that it was a fallacy to contend that man was ordained, unless he limited his offspring, to increase more rapidly than he had ability to find the means of subsistence in a corresponding ratio. It was an error to maintain that the production of the necessary means of subsistence became more difficult as the population increased.

The contrary would be the case if education and intelligence to produce co-operatively also increased. More shepherds could live on a square mile, and live more luxuriously, than rambling hunters, and more farmers than shepherds. The greater the population the greater the control which man had over nature, and the more easily could the means of subsistence be obtained. Through increased population, there was not a movement from better to worse soils, as taught by Malthus and Ricardo, but the reverse was the case; for just as progress was made from bad to better methods, so also there was a movement from bad to better soils, from which the means of subsistence could be obtained with less cost and in greater abundance. In opposition to Smith and Ricardo, he taught that rent was nothing else than interest on capital. The land in many under-populated regions was to be had for nothing, and for which no rent was paid. When land through centuries of expenditure of capital became rich, then a high rent was naturally demanded for the use of the soil; but this interest was by no means an ample return for the capital sunk in the soil at the current rate during the given period. In nature there was no incongruity between the wants of the growing population and the possibility of increasing the means of subsistence, so that there could be no discord amongst the interests of the landlords, the capitalists, and the labourers. Measured by the amount, wages and capital, as well as rent, grew with the increase of wealth; but reckoned in percentage of the total amount of the national production, wages were ever on the increase, while rent and interest sank deeper and deeper. An ever-increasing portion of the products went to labour; an ever-diminishing portion to possession.

In France, the theories of Adam Smith made a deep impression upon the minds of the economists Say and Sismondi, while the concord theories of Carey impressed the economist Bastiat, who, however, was an extreme Free-trader. In Germany, Smith made little impression at first, but latterly his theories caused a good deal of criticism, and received some acceptation in modified forms; but the German economist, Hagen, arrived at the same conclusions as those entertained by Carey's school of concord.

The lack of scientific method amongst the economic theorists is now made plain. The ancient Greek philosophers, who are the divinities of modern speculative thinkers, recognised agriculture to be the basis of national wealth, and the same view was held by the Physiocrats. Instead of solving this dualism of economicism and agricolism, or making any serious attempts to do so, the economists split up their theories into a number of petty, interior dualisms—*e.g.*, free-trade and protection, concord and discord, wage-fund and anti-wage-fund, labour *versus* money as the standard of value, &c.—and during all this time agriculture was allowed to shift for itself. The minor dualisms can never be conciliated until the ulterior dualism is solved, and even this may be of no avail so long as the ultimate dualism remains unsolved. No solution having been offered by the speculative thinkers, it was left to the operations of the laws of nature, and, in the next economic wave, we find the results of the dualism being coerced to its bold extremity, just as we traced the reactionary process in our religio-philosophic inquiry.

Economic agencies forced abundance into the hands of the oligarchic few, while it robbed the democratic many of adequate necessaries. In order to restore the equilibrium, a new economic force appeared, wrought into a system under the name of *Scientific Socialism.* This school had three distinguished founders, namely, Rodbertus, Marx, and Lassalle—men of profound learning in jurisprudence and economics, and of extraordinary force of character, whose aim was to move and lift the masses. The two last named were also great philosophers, and being of Jewish extraction, naturally felt the pangs of oppression very keenly, the Jews being a race who for many centuries have suffered acutely for the shocking crime of adhering to the worship of their remotest ancestors, and for the profound subjectivity of the Semitic character. This school has now won a page in all the leading works on political economy, and its platform is, briefly, as follows :—

Wealth consisted of values (utilities). Only those goods had value which were the products of labour ; and the quantity of labour, at the given term and under the given methods, determined the value of the product. In the economic sense, only those people were producers who, by the direct application of their muscular and nervous powers, aided in producing values. Only those who actually laboured were producers.

Landlords and capitalists—all those who lived from rents, loans, stocks, and other incomes—were not producers; they did not work, did not produce, but consumed, values—the products of other people's labour. Artists, scholars, poets, doctors, judges, &c., were in reality not producers, but were useful members of society, and deserved to be rewarded by the producers. So long as the soil and other sources of production remained private property, and wages were regulated by free competition, the labourers could not receive more from the total produce of their labour than was necessary to live, work, and bring up their families for work. Under the control of these agencies panics must prevail, and a portion of the labouring classes must be without food and work, despite the false theories with reference to over-population as taught by the Malthusians. According to the principles of right and reason, and in sympathy with sound economics, land and capital should, in some way, become the property of the State (society, the producers)—such as by expropriation with little or no compensation. State officers should be appointed to conduct all the industrial enterprises in such a manner as post-offices, railways, joint-stock companies, co-operative associations, &c., were now conducted. There would then be no rents or free incomes, and the producer would be rewarded by the receipt of such a proportion of the total products as represented the produce of his labour.

The founding and developing of this school tend to subjugate the interior and inferior dualisms of our orthodox economists, and of necessity a new dualism must be created, the opposing force now being *Individualism*, the great champion being Mr. Herbert Spencer, our evolutionary philosopher. The policy of the Individualists is to thwart the progress of Socialism. They give the individual the utmost freedom, consistent with equal freedom on the part of others, for the exercise of his faculties, and limit the functions of the State to the protection of life, liberty, and property, and the administration of justice. They are practically a property defence association, the members being composed of men of property. In England, the Socialists belong mainly to the masses, excepting some learned bodies such as the Positivists; but on the Continent, the school includes many professional men—"Socialists of the Chair"—and the Socialistic tendency in all countries is strongly atheistic, while the leanings of the Individualists are theistic, although not always ecclesiastic.

An exception to the more orthodox Socialistic school, however, is found in *Christian Socialism*, a sect largely influenced by two English clergymen, F. D. Maurice and the late Charles Kingsley, who, according to Rev. H. C. Shuttleworth, "claims to be the result of applying Christ's teaching to national, social, and commercial life, and not merely to personal conduct. Those who hold this view maintain that Christ said little as to a future state, but much of bettering the conditions of life in this world. They point out that He consistently placed the community before the individual, and taught that the foundation of society is brotherhood, not competition for profit, as now with us. Christian Socialists adopt that name · because they believe that a really Christian society must be what is called socialistic." The economic theories of the Christian Socialists are mainly those of Henry George, a Christian economist, who, elaborating the principle laid down by Herbert Spencer, to the effect that the land belongs to the people, only goes so far as to nationalise the land, and, therefore, not the products of the land. Mr. George maintains that when free access to the land can be had wages must rise, because the labouring classes will then resort to agriculture if they are dissatisfied with the wages offered them in industrial employments, thereby lessening the competition. The Fabian Society was founded (1884) for the purpose of spreading Socialism amongst the educated classes. The main difference between the Socialists and the Georgists is, that the programme of the former is, first of all, to nationalise the capital, the land being a secondary consideration so far as the labouring classes are concerned, and they further believe that land nationalisation alone will not materially, if at all, improve the condition of the labourers; while the supporters of Henry George generally believe that the nationalisation of the land is an adequate remedy for all our social ills, although some believe that this step would, sooner or later, be followed by the nationalisation of all property, as advocated by the Socialists. Mr. George fails to point out whether he regards agriculture or trade as the basis of national prosperity, but the inference which I draw from his writings is that his scheme is for the purpose of advancing the interests of economicism. .

Neither Individualism nor Socialism, including Georgism, is one step in advance towards the solution of the social problem. Existing forces will unquestionably drive us into Socialism, and it may be followed by a period of peace and contentment, but no permanent solution can take place until the agricolo-economic dualism is subjugated.

The writings of Auguste Comte (1798–1857), the founder of Positive Philosophy, have given origin to another system, which has received its highest development in Germany, namely, the *Historico-Ethical School*. In proportion to the decline of the old orthodox theories which ruled from Smith to Cairnes, this school has been making its influence felt in England. Its disciples have adopted a new method of investigation, mainly through historical comparison. It gives social and ethical duties precedence over individual rights, and makes society subject to natural laws. Its teachings are substantially as follows :—

It was a fallacy to suppose that a social system could be grounded which was equally suitable for all times, all nations, and all grades of culture. Any system, regarded as efficient, must harmonise with the requirements of the age, nation, and educational standard. Society was not merely a mass of individuals and families living at a given period, but also embraced many generations following each other—a perpetually organised community, or an organism whose members were individuals and families. The State was not a mere medium, institution, or bond existing for the protection of the interests and liberties of the individual, but, moreover, an organisation through which the people, the nation, formed a totality throughout all time, and which exerted itself in the service of the highest moral purposes, including the security of the individual against aggression in the exercise of his free moral development. In a healthily developed State, the economic adjustments should neither be entirely in the hands of the Government (Socialism), nor left exclusively to the devices of the individual (Physiocratism). The nice medium between the State and the individual could not be theoretically determined once for all : the varied requirements of the times were the determinator. It was a fallacy to maintain that in economic and kindred affairs only the material and selfish interests were to be regarded, and that the moral standpoint was to be left entirely to the conscience of the individual. Indeed, those laws which pertained to property and intercourse should be based on a moral standard. Only for the sake of freedom, without which no moral development was possible, and touching the dangers which attended too great a restriction of freedom, must much objectionable conduct be left to the conscience of the individual and to the morality of social restraint. But when danger was near, which threatened to demoralise many branches of industry, or to extinguish national life, then it became the duty and consequent right of the State to grasp with iron hand and regain the labour, the property, the commerce, the sovereignty.[1]

Much as I admire this historic school and the many able writers which gave it birth and guided its development, yet I must confess—and it pains me to be obliged to do so—that it is destined to enjoy a short career unless it extricates itself out of the economic category. It has earnestly espoused the cause of humanity, but so long as it confines its sphere to the conciliation of the Socialistico-Individualistic dualism, its brilliant anticipations can never be realised. This school is specially noteworthy on the ground that many competent authorities entertain the hope that it is destined to dethrone the economic orthodoxy which still holds sway.

This summary would not be complete without noticing, finally, *W. Stanley Jevons* (1835-1882), and his relation to the various schools of economics. Apart from his titles, LL.D., M.A., F.R.S., he writes as a scientific authority, although his native sphere seems to be philosophy and logic. I mention this fact because, as I have shown, there has been a great dearth of scientific talent amongst previous economists. As a mathematician of a high order, he laboured hard to

---

[1] Since gathering the materials for this chapter, chiefly from various English writers, I accidentally fell across an estimable little work entitled "*Elemente der Volkswirtschaftslehre*," by Dr. W. Neurath (published by Julius Klinkhardt, Leipzig, 1887), from which I have made a free translation of the above paragraph. Indeed, I am also indebted to the same author for much of the summary in previous portions of this chapter. The English mind is economic, while the German mind is scientific : this explains the reason why English writers cannot be understood except through the interpretation of German critics and reviewers.

reduce economics to mathematical formulas, and yet he was also a faithful worker in the cause of the masses, which gave his mind a Socialistic bent. He endeavoured to bring economics into closer relation with the physical sciences. He says : "I have no hesitation in accepting the Utilitarian theory of morals which does uphold the effect upon the happiness of mankind as the criterion of what is right and wrong." Accordingly, he treats his system of economics as a " Calculus of Pleasure and Pain," and then attempts to reduce the vibrations of pleasure and pain to mathematical rules. This attempt must prove fruitless from the fact that the basis is an abstraction which cannot be subjected to definition ; and, besides, few people agree that the Utilitarian theories are the best in the world. Indeed, this basis does not shift that of the old orthodox school, namely, human desire ; for everybody seems to desire such things as afford pleasure (real or imaginary) and mitigate pain. Man, as an economic animal, realises pleasure in making or hoarding up wealth. Is this our highest ideal of happiness? Again, he is very unscientific when he says that "the present chaotic state of economics arises from the confusing together of several branches of knowledge : subdivision is the remedy." He has carried out this idea by burdening his mind with many of the petty little dualisms which I have pointed out, instead of taking the broad scientific view. He attributes the periodically recurring commercial crises to the sun-spots. This seems like an attitude of despair. In later years he seems to have courted the historical school—at least he saw the importance of utilising historical methods. He may fairly, I think, be classed as marking the transition stage in political economy. It is unquestionable that the old orthodox school cannot give birth to any more distinguished economists.

My summary is now completed, and although it embraces only the most glaring differences of opinion, yet no apology, I think, will be demanded should I seek a broader basis of inquiry. Economics, as we have seen, is as old as philosophy, and the dangers of economicism were foretold by the great philosophers of ancient Greece, as well as by the Church during the Middle Ages, and yet it enjoyed its high state of development almost entirely through the instrumentality of our philosophers and theologians during the past century and a half.

It is here in place to touch upon the relation between the ulterior and the ultimate dualism—between the agricolo-economic and subjectivity *versus* objectivity—although the development of this relation belongs to succeeding pages. It will be remembered that the religio-philosophic dualism resolved itself into subjectivity and objectivity. We find precisely the same law in agricolo-economicism, although education, individual temperament, and other forces have been the means of rendering it more or less obscure. The philosopher thinks that he can transcend the limits of nature or experience and penetrate the absolute, and the religionist feels that he is under the sway of higher powers which he must obey and which must be made known to him through revelation. In short, the former bases his knowledge upon abstract reasoning; the latter upon Authority. So it is in the material world : there are some who, by their way of thinking, regard man's mission to be the exercise of his powers against nature, while others believe that natural laws should be permitted to enjoy their course, man's duty being to obey them. Reaction from objective extremes in philosophy led to Sophism, Cynicism, and Stoicism; and in economics the reaction from the objective extremes of Mercantilism led to Physiocratism and Free-trade. In every dualism we find a subjective and an objective element, and sometimes, even in the same individual or organisation, we find the objective basis predominating in the moral, and the subjective in the material sphere, and *vice versâ*. Socialism, for example, is subjective from the standpoint of humanity, and objective from the standpoint of economicism. To trace historically the waves of subjectivity and objectivity, and to solve this dualism by finding which one is the basis of the other, is the scope of this work, and herein resides the solution of the social problem. The dualism of objectivity and subjectivity may be expressed in other language, namely, man and nature. The question now resolves itself into this : Is man subject to nature, or nature subject to man ? In a scientific inquiry, all supernatural agencies must be eliminated; for if man is impregnated with supernatural powers, the problem is already solved, nature then being under his dominion, and he is not amenable to her laws.

Having shown the insufficiency of political economy from the negative point of view—and I have merely pointed out the most glaring contradictions—let us now discuss a few of the positive weaknesses. It is not necessary to say much about the orthodox school, for it has committed suicide; and with reference to Socialism and Individualism, which are political as well as economic forces, it is first necessary to examine the character of the State. With reference to our financial dualism, monometallism and bimetallism, which is a material issue in the solution of the social problem, so great is the necessity for establishing scientific standards of value that I shall devote a whole chapter to the subject. It is merely necessary to remark here that our financial economists are striving to base our monetary system on the soundest principles of political economy, and that, as these "soundest principles" are fundamentally false, our financial arrange. ments must suffer the same fate, and totter into the dark abyss as a component part of the same fabric. Wealth, money, capital, labour, wages, interest, rent—all stand or fall together; they are timbers of the same structure—economicism.

The early economic philosophers who held that labour was the measure of value showed a great weakness in not establishing a labour-standard, instead of adhering to the standard of money. This failing is specially applicable to Adam Smith, the Christian philosopher, who drew such a sad picture of distress and suffering amongst the labouring classes— the people who produced all the wealth and received empty stomachs for their wages. It is to be presumed that if labour were made the standard of value, the labourer would receive at least a sufficient percentage of the products of his labour to enable him to live and grow. Locke said that the products of the earth derived "almost all their value" through labour. The Christians and sceptics were a unit on this question; for Hobbes believed that "plenty" depended on the labour of man.   The Socialists, who feel so confidently on this point, are guilty of the same default.   Jevons says that "value depends entirely upon utility"—therefore not upon labour.

Wealth, one would naturally suppose, is composed of values—"plenty," as Hobbes puts it.   This, however, is not the case, and on this point all economists are a unit.   Senior, whose definition has never been disputed, confines it strictly to "those things which, directly or indirectly, are made the subjects of purchase and sale, of letting and hiring," and explains that such things must therefore, in the first place, possess "utility"—be capable of affording pleasure or preventing pain.   Secondly, they must be *limited in supply* (*i.e.*, scarce !) — for they would not constitute any portion of wealth if they could be had for nothing.   Thirdly, "nothing is wealth that is not capable of appropriation"—*e.g.*, fine weather.   And, fourthly, it must therefore be directly or indirectly transferable.   Many of our orthodox economists neglect to define their ideas of wealth; but it has never been disputed that human desire is the basis.   In political economy, all kinds of desires are equal, and equally moral so long as they produce equal increments of wealth (scarcity ?).   Jevons, who appears to have established a different basis, repudiates desire and substitutes pleasure—a foundation equally abstract in its character—and he has failed to give us a list of those commodities which afford us pleasure, but which we do not desire, or those which we desire, but which do not afford us pleasure.   By drawing this line, his shift of basis would be more readily comprehended.   He may have considered it easier to invent an instrument for measuring our "vibrations of pleasure" than those of our desires.   Henry George contends that nothing can be wealth which is freely supplied by nature (all natural opportunities).   He excludes such things the destruction of which would not decrease the general wealth—such as "bonds, mortgages, promissory notes, bank bills, or other stipulations for the transfer of wealth."   Slaves, in his economics, are also excluded from the category of wealth.   The various definitions of that portion of wealth called capital are too numerous and complicated to be noticed here.

Arriving at the conclusion even more clearly, Professor Jevons announced "that the only hope of attaining a true system of economics is to fling aside, once and for ever, the mazy and preposterous assumptions of the Ricardian school." Mill, who showed that political economy was based on hypothesis, and dealt merely with "economic men," "money-making animals," which view was also entertained by Cairnes, says: "There is nothing in the law of value which remains for the present or any other writer to clear up; the theory of the subject is complete." Senior, in 1826 and again in 1852, spoke of political economy as being in an imperfect state of development; and M'Culloch stated that the differences amongst the most eminent professors created a distrust in its conclusions, although he had confidence that the errors with which the science (!) was infected were rapidly disappearing. Many years ago, Torrence prophesied that in twenty years there would scarcely exist any doubt respecting any of its moral conclusions; and Cobden confidently expected that his free-trade theories would speedily spread over the whole civilised world.

If our economists could agree in the definition of their science, or decide whether it is a science or an art, or could tell us whether the inductive or the deductive method of investigation should be employed, or give us a vague idea of its scope, the dawn of better days might yet be hoped for respecting economic orthodoxy. Professor Cairnes burdens us with the following definition : " The science which, accepting as ultimate facts the principles of human nature, and the physical laws of the external world, as well as the conditions, political and social, of the several communities of men, investigates the laws of the production and distribution of wealth, which result from their combined operation ; or thus :—The science which traces the phenomena of the production and distribution of wealth up to their causes in the principles of human nature, and the laws and events, physical, political, and social, of the external world." The weakness of this definition is that the author failed to inform us whether he meant the human nature as moulded by the theories of economic orthodoxy, or those which would have been moulded had some other school, say the Physiocrats, swayed the public mind—or under an economic rule by which plenty, instead of scarcity, was made the basis of wealth. It must not be inferred that I am the author of the wild abstraction that a commodity must be scarce before it can be classed as any portion of wealth. This is an important feature in our inquiry, and in order to avoid misconception, I quote the authority [1] of one of the highest institutions in our land, as follows: "Articles which form part of wealth must be *limited in supply;*" and the author follows the Utilitarian philosophers in attributing *utility* to those articles which, directly or indirectly, produce pleasure or prevent pain. The University economists also consider that commercial crises may be traced to the sun-spots. This superstition is weak from the standpoint that its worshippers fail to inform us what effects the sun-spots produced on our earth before the days of economic rule.

The true inwardness of economics may be inferred from its assumptions. Let me quote the "first assumption" from the same author, namely: "That the industrial actions of men are determined by the desire of obtaining as much wealth as possible with the least possible exertion." Exactly so; and this is just the very reason why we are obliged to struggle so hard for the means of subsistence—because it is our *desire* to get much and do little. The economist creates human nature, and then assumes how it will act under given conditions. He creates wealth by making its commodities scarce, and then assumes that it is human nature for a hungry man to steal. He maintains, moreover, that he has nothing to do with moral problems, that political economy has only to do with the material well-being of mankind, that it instructs our politicians how to make the poor as few as possible, and how everybody may be well paid for his work—in short, how a nation may be able to enjoy the largest possible quantity of scarce commodities. If moral problems have nothing to do with economics, then happiness or desire, as its foundation,

---

[1] "The University Economics," an elementary text-book on Political Economy, by E. C. K. Gonner, M.A. (Oxford). Published 1888. The italics are the author's, not mine.

must be rejected; for even in the palmy days of Greek philosophy, before the era of economic rule, not to speak of modern Utilitarianism, happiness played a leading *rôle* in ethical philosophy. This "first assumption," of which we have spoken, is a dangerous theory from another point of view. It assumes two propositions: (1) that men like wealth, and (2) they· dislike labour. Why do men like wealth? Because they are economic beings. Why do they dislike labour? Because they do not receive the products of their labour. When men are engaged in robbing their fellows, and hoarding up scarce commodities for themselves and their "heirs and assigns for ever," they do not then dislike labour. Such men do not strike for eight hours a day. Does the scientist or the philanthropist—men who are engaged in elevating their fellows and pleading the cause of humanity—dislike labour? Labours of love and justice are foreign to the economic mind. There is no need of a taskmaster to urge economic men on during the pleasures of their midnight employments, and yet we are told in tones of the greatest coolness and most owlish gravity that those who wear their lives away to support them in all manner of licentiousness dislike labour.

Logicians have played a prominent part in the moulding of economic theories, and I most respectfully invite them to point out the fallacy in the following proposition :—

Commodities must be scarce before they can be classed as wealth : therefore the scarcer they become the more abundant grows the wealth.

If there is any limit to these extremes, then I want to know what ratio between abundance and scarcity is conducive to the maximum quantity of wealth. When a commodity becomes so abundant that nobody desires it, it ceases to be wealth, from which it must surely follow that when everybody desires it, and is ready to sacrifice his life and all his possessions for it, the wealth, so far as that commodity is concerned, must reach its maximum. If clothes grew on our backs, and roasted chickens flew into our mouths—in short, were all the objects of our desire to become as abundant as air—there would be no wealth, and consequently everybody must starve, because over-production brings such direful calamities. If this is not a practical illustration, let us take another. During the past twenty years, £6,000,000,000 have been added to the capital of the United States, while the means of subsistence, taking wheat as the representative, have decreased from 25–30 bushels per acre to 12–15 bushels for the same period. Half a century ago 35–40 bushels were readily obtained; and yet the land has increased in value, the virgin soil having been obtained either as free homesteads or for the low price of 3–5 shillings per acre, while it now brings £10–15 per acre. This increase in economic value cannot be due to the improvements, for the wooded and un-improved portions of the same land bring as much as, and sometimes more than, the cultivated and improved areas. Carrying these figures to their logical conclusions, the time must come when there will not be sufficient fertility in the soil to support a grasshopper, while the wealth of the great Union, in bonds and mortgages, will have been so enormously increased that there can be no possibility of over-population, even though all the inhabitants be kept exclusively for breeding purposes. If this illustration is not satisfactory, let us take one in our own country. According to the Report of the Royal Commission on the Depression of Trade, the capitalised loss by owners of agricultural land and their tenants in 1885 amounted to £740,000,000. The land itself has unquestionably increased in fertility, if not also in pro-ductiveness, owing to the immense quantities of cattle-food purchased abroad and fed on the land, and also owing to the great quantities of commercial fertilisers broadcasted over our pastures and cultivated fields. Reversing the American con-dition, let us now drive these figures to their logical conclu-sions. The time must come when the capitalised loss will be an inconceivable amount, while the land will be so rich in the means of subsistence that, like the blackberries growing in our hedges, the crops will become so abundant that they will bring no money; agricultural wealth will then be totally annihilated, and consequently our posterity is doomed to starvation, because they will have no money to live on. These illustrations are ample to show that economic wealth is based upon scarcity, not upon abundance, and it is a negation as well as an abstraction.

The economist can, of course, explain this situation by the theories of monopoly, and this is just the battlefield into which I have been attempting to pursue him. Here is the ground on which the battle must be fought. It is an easy business for our politicians to create economic wealth. By selling the Atlantic Ocean, or our streams, to some enterprising corporation, our economic wealth could be enormously increased, for the scheme would give employment to an enormous number of our unemployed labourers, and many of our respectable and highly educated people would be required to collect rents and other revenues, to act as watchmen, &c., and then just imagine the number of buildings that would be required to line our shores. If our nation would not then be wealthy enough, an attempt might be made to get our air monopolised, because it would require an enormous quantity of labour, machinery, &c., to pump the air out of people's houses and fields and preserve it in reservoirs made for that purpose. It would also be an excellent plan to pass stringent laws to compel us to breathe nothing but pure Italian air imported in bladders, which would give respectable employment to our lawyers and judges, not to mention the encouragement given to trade. A heavy fine should also be imposed on all who attempted to change the fashions, because the old commodities, being no longer desired, would cease to be wealth. Besides, a bounty should be given to all inventors of new fashions, because new things always bring much money, thus increasing the national wealth. History proves that war is a great wealth-creating agency, for it destroys so much property, thus making it scarce, and it restores confidence in trade. Slaughtering our brothers on the battlefield appears to be the most effective method of restoring public confidence and strengthening our temporal powers. Would the maximum of wealth be attained when every commodity was destroyed, or when war raged until every human being was wiped from the face of the earth? Would confidence in trade and in our fellow-beings be completely restored, once and for ever, when the fury of war converted mankind into messes for vultures and buzzards? If this extremity will not do, then why not try the other, namely, the creation of maximum wealth by everlasting peace, or say what ratio between peace and war is most conducive to the production of maximum wealth? How would it do to conciliate the dualism by inaugurating a peace footing in times of war?—a war footing in times of peace? To these questions, O Political Economy! I pause for a reply.

The fact that scarcity is the basis of economic wealth is further proved by the action of our monopolists. The very idea of monopoly is the power of making the monopolised commodity scarce, for the greater the abundance the farther it is removed from the catégory of wealth, and the degree of scarcity is measured by popular indignation and resentment. Nobody wants a monopoly of a commodity which exists in quantities above the range of human desire, present or prospective. It makes little or no difference whether the commodity be a necessary or a luxury, for when a desire becomes fashionable, it must be appeased at all hazards. Indeed, the scarcer the commodity, the more fashionable it becomes. It becomes fashionable by virtue of its scarcity.

The whole secret, as we shall see, may be summed up in a few words, namely, that natural law refuses to twist itself into conformity with economic orthodoxy—that there is a trifle too much objectivity in the minds of our abstract theorists. Economicism has failed to create the laws of nature. Our labouring classes have discovered by hard experience, not by theory, that even Providence Himself refuses to reconcile economic wealth with hungry stomachs, even through the intercession of philosophers and priests. So long as these functionaries can convince the masses that their theories are sound, and that any violation of them will be attended with painful consequences, peace and starvation will flourish side by side ; but war is the inevitable effect of driving so violent a dualism to its dire extremity.

CHAPTER IV.

THE THEORIES OF POPULATION.

THE population question is the root of the social problem, and it cannot therefore be so summarily treated as other economic theories. Although it properly belongs to the last chapter, yet a special chapter is required even for the faintest amount of justice to the subject. The quantity of economic wealth cannot be determined until it is ascertained whether or not the given area is over-populated, and the extent of the population is also the basis of the scientific wealth. I shall sum up briefly the weaknesses of economic theories, thus gaining space for the positive method of demonstration, instead of placing the main reliance on negative proofs.

In the population controversy, the disputants may be divided into two classes: (1) the Malthusians, who think the world, or some given area, is over-populated; and (2) the anti-Malthusians, who deny the correctness of this statement. It was the former policy of our rulers to encourage population, their rents thereby becoming increased, which was in sympathy with national prosperity. Nowadays, however, this order is reversed, the tenants not being able or willing to pay the high rents thus created, and, their political power also increasing, the cry of over-population resounds from the classes who still enjoy their inning. The labouring classes, on the other hand, maintain that there would be no over-population if our national wealth were more evenly distributed—that monopoly is the source of the prevailing misery in all civilised countries. In this chapter I shall point out the number of inhabitants which the United Kingdom can support in food, leaving the question of monopoly for later discussion.

Axiom II.—*The population which can be sustained is based upon the average area of land required to support each individual.*

Before entering into the calculations required to solve this problem, let us consider a few of the weak points in the Malthusian theories.

If civilised countries are over-populated, what portion of the population is to be removed? If it is to be the lower classes, which is the lower class, those who produce the commodities which support others in licentiousness, luxury, and ease, or those who live without compunction on the life-blood of the producers? Which of these classes has the greatest right to live or remain in their own country? If the poor but educated and refined are to be removed, should the first victims be selected from the politicians, the economists, the philosophers, or the priests? Should it be the unemployed poor, or the unemployed rich? If the hours are to be legally restricted, should it be the hours of labour or the hours of idleness? Should the hours of labour be restricted so long as people are famishing for want of more wealth, or should idlers be prevented from working for fear of over-production? If those who cannot make a living are to be removed, then who is to get the first chance—who is to get the first monopoly— the strong or the weak? The weakness of the situation is, that there would be a great variety of opinion on these questions, and also with reference to the methods of removal. For example, one of the most effective measures for the removal of the surplus population in all civilised countries is through dissipation and debauchery. Under this system, whose liberties are to be imperilled by laws for the enforcement of morality? If war is to be proclaimed as the universal panacea, should those who wage the wars do the fighting, bleeding, and dying, or those whose voice is for universal peace? Are the poor to fall on the field of battle and the rich to escape unscathed? If the remedy is to be in the delaying of marriage until the doomed persons obtain the visible means of subsistence, will the Malthusians establish and maintain a fund for procuring divorces and supporting the families of those misfortunate speculators who to-day are millionaires and to-morrow tramps? If, in order to enhance the dignity of labour, mental instead of physical labour be introduced into our prisons, should the most atrocious criminals be sentenced to study political economy, philosophy, or theology? If natural law is to be the judge in these matters, is the man of the highest social standing the first to flee the flame-devouring deck? If that most impartial of all judges, Blackdeath, were again to exhale his pestilential breath over our sovereign city, would the poor be taken and the rich left?

The area of land drawn on for a person's subsistence may be divided into four classes, namely: 1. That required to supply him with food. 2. The area drawn on for beverages (wine, beer, brandy, tea, coffee, cocoa). 3. For external shelter and warmth (clothing, houses, fuel, light). 4. Medicines (under this heading tobacco, opium, &c., must be classed, in addition to ordinary drugs). In this chapter, I am obliged to confine the discussion to the area of land required for food, there being other subjects to be discussed before we can determine the area required for clothing, beverages, &c. With reference to mineral ores, which are the raw material mostly used for tools, implements, machinery, &c., employed in the manufacture of our food, clothing, and other articles of consumption, the principle is the same as that applied to the fertility of the soil, what concerns us being that in both instances (1) exhaustion takes place, and (2) labour is required to manufacture the raw material into the finished product. The soil may be regarded as an exhaustible mine, the elements of fertility being the raw material of our food, clothing, beverages, tobacco, &c.

Some agricultural plants contain a larger percentage of nutriment than others, and some yield heavier crops, so that the class of plant, as well as the quantities consumed, must be determined before we can calculate the area of land drawn on for the support of the consumer. Again, a large percentage of our people consume animal food, in which case the quantity of land devoted to the support of the domestic animal must be ascertained, also the quantity of meat consumed. Let us assume, meanwhile, that the consumer will require varying quantities of land for his support depending upon whether his ration is drawn, mainly or wholly, from the vegetable kingdom, or from the flesh of domestic animals, mainly or wholly. In adherence to the scientific method of inquiry, we must now ascertain whether man is a carnivorous or an herbivorous animal, or whether or not he is omnivorous. It will not do to conclude that he is omnivorous because he consumes mixed rations, for I would then also be obliged to call him an alcohol animal because he imbibes alcoholic beverages, or a tobacco animal because he consumes that weed. Besides, there are millions of men who do not consume flesh, and millions more who consume no vegetable food; so that if the question is to be decided by the habits of our race, we are carnivorous, herbivorous, and omnivorous all mixed up, or more or less combined. Should we find that man, although originally herbivorous, has, through ages addicted to flesh-eating, evolved into a carnivore, or at least into an omnivore, then it will be my duty to point out the corresponding changes of structure and function, and also to show the advantages to be reaped by such a process of evolution.

This question, however, is tripartite rather than dualistic, for there are three kingdoms of nature, animal, vegetable, and mineral, and as there are some people who believe that man can live partially upon the mineral kingdom, the problem is not confined to vegetarianism *versus* carnivorism. It is well known that man, or any other animal, cannot live without organised minerals (ash, salts), which are found in all plants and in all animals, but this is not the question. It is contended that when mineral salts are deficient in our rations—usually caused by artificial methods of preparation, such as cooking and milling—the deficiency may be restored by drawing the lacking salts directly from the mineral kingdom before being organised into plant tissue. Accordingly, some people take drugs, mineral waters, and especially common salt, for the purpose of supplying any saline compounds that may be wanting in their daily rations. Granted that some people's food may be flavoured by this treatment, namely, those who have blunted their tastes for natural flavours through the use of spices, condiments, &c., or those who consume unnatural food, or such food as has lost its flavour through partial decomposition or mustiness, yet these practices have nothing to do with the discussion of the nutritive, or non-nutritive, properties of unorganised minerals. If these salts possess medicinal properties, and are taken as medicines by those who require drugs in all their rations, the question then assumes quite another aspect. Now, if common salt must be supplied to make up for the deficiency of saline matters in our rations, the inference is either (1), that salt is the only ash constituent removed (or rather removed below the point of sufficiency) ; or (2), that it can be substituted for those other salts which have also been removed. The salts in plants which are absolutely necessary for their existence are potash, lime, magnesia, iron, phosphoric acid, and sulphuric acid. All plants also contain chloride of sodium (common salt), but they can thrive without it, although animals cannot. The question which now arises is this : If vegetable or animal food were deprived of all its saline constituents, could the loss be supplied from the unorganised mineral kingdom ? The affirmative of this question has never been maintained, for it would lead to the conclusion that animals could live on chemical fertilisers, or on soil. As nobody has ever been foolish enough to attempt, or even suggest, such an experiment, we must conclude that unorganised minerals possess no nutritive value so far as animals are concerned, and that the superstition originated in a lack of scientific knowledge. As the saline constituents of plants are required for the building up of animal tissues, no one salt being capable of substitution for another, the scientific conclusion is, that if, say, common salt is deficient in our rations, the only remedy is the utilisation of this mineral as an agricultural fertiliser, so as to become first organised into plant life. This question, however, is not of vital importance just here ; but as it will be of great concern in succeeding pages, I place special stress upon it in this place.

Having eliminated the mineral kingdom from this section of our inquiry, it remains now to consider whether our rations should be drawn from animal or vegetable nature.

Although few people contend that our rations should be

drawn entirely from the animal kingdom, yet pure vegetarianism is rapidly spreading, and its doctrines have a scientific basis. A weakness amongst the defenders of rations composed wholly or partly of flesh is, that they have not been able to show whether the flesh should be from the carcasses of animals which are herbivorous, omnivorous, or carnivorous, or whether the process of our evolution is from herbivora to carnivora through the medium of rations composed, wholly or partly, of the flesh of omnivorous animals. This point also gives rise to another question, namely, Is such an evolution a law of nature? If not, it being through our own free-will, what advantages can we enjoy by such an evolutionary process? These weaknesses suggest another line of inquiry. Accordingly, I shall start out from the vegetarian basis, and shall ascertain the area of land required to support an individual on a vegetarian diet, then also on a flesh diet, so that the defenders of the mixed rations will have an opportunity of accepting their own medium between these two extremes. Before accepting this basis, however, we must examine whether man is an herbivorous animal.

That man is structurally related to other mammals is now recognised by all biologists of repute. Darwin says:[1] "It is notorious that man is constructed in the same general type or model as other mammals. All the bones in his skeleton can be compared with corresponding bones in a monkey, bat, or seal. So it is with his muscles, nerves, blood-vessels, and internal viscera. The brain, the most important of all the organs, follows the same law, as shown by Huxley and other anatomists. Bischoff, who is a hostile witness, admits that every chief fissure and fold in the brain of man has its analogy in that of the orang; but he adds that at no period of development do the brains perfectly agree, nor could perfect agreement be expected, for otherwise their mental powers would have been the same. . . . Man is liable to receive from the lower animals, and to communicate to them, certain disorders, as hydrophobia, variola, the glanders, syphilis, cholera, herpes, &c.; and this fact proves the close similarity of their tissues and blood, both in minute structure

[1] Descent of Man, vol. i. p. 7.

and composition, far more plainly than does their comparison under the best microscope, or by the aid of the best chemical analysis. Monkeys are liable to many of the same non-contagious diseases as we are: thus Rengger, who carefully observed for a long time the *Cebus Azaræ* in its native land, found it liable to catarrh, with the usual symptoms, and which, when often recurrent, led to consumption. These monkeys also suffered from apoplexy, inflammation of the bowels, and cataract in the eye. The younger ones, when shedding their milk-teeth, often died from fever. Medicines produced the same effect on them as on us. Many kinds of monkeys have a strong taste for tea, coffee, and spirituous liquors : they will also, as I have myself seen, smoke tobacco with pleasure. Brehm asserts that the natives of North-eastern Africa catch the wild baboons by exposing vessels of strong beer, by which they are made drunk."

Darwin regards these facts as proving the similarity of the nerves of taste in the monkey and man, and " how similarly their whole nervous system is affected." He then continues the similarity by reference to parasites, the course of disease, reparation of wounds, reproduction of the species, the acts of courtship by the male, nurturing the young, &c.[1]

Cuvier, one of the greatest of comparative anatomists, says : " The structure of the human body points in every essential particular to a vegetarian dietary. Man appears to be con-structed very largely for eating fruits, roots, and other juicy parts of plants. His hands impart to him the facility to pluck them, but his short and moderately strong jaws, as well as his incisor and molar teeth, are unsuitable for grazing or eating flesh, if he did not prepare his food by cooking."

Professor Gassendi says : " According to the original and

---

[1] Since Darwin's time, all searching investigations have proved man's genealogical relation to the lower animals, and amongst the German scientists, who are slow to accept theories without substantial proof, are ardent supporters of the doctrine of evolution. For the latest proof, I refer the reader to *Der Bau des Menschen als Zeugniss seiner Vergangenheit*, by Dr. R. Wiederschein (published by the Akademische Verlagsbuch-handlung, Freiburg (1887). This distinguished author has collected a mass of evidence with reference to man's process of evolution. French scientists have come to the same conclusions.

unvitiated arrangement of our nature, our teeth are not suited for masticating flesh, but only food from plants."

Professor Huxley says: "Whatever part of the animal structure, whether series of muscles or viscera, we select as a basis of comparison, the result is the same. The lower monkeys and the gorilla differ more widely than do the gorilla and man."

Häckel says: "One may compare any parts of the body whatever, and it will be found that man is nearer the highest apes than the latter are to the lowest apes. But the highest apes are purely vegetarian; therefore also man. The zoologist is compelled, whether it is agreeable to him or not, to rank man within the order of the true ape (Simæ)."

If, after all this evidence—and much more could be educed—any doubt remains, Professor Huxley, in his classification of mammals by their placental structure, has effectually quelled all opposition, the placenta being regarded as the finest mark of distinction between carnivorous and herbivorous animals. The teeth, in man and the ape, are identical in structure, number, and position. All these resemblances, however, are not identical with reference to strength; for our teeth have become much weaker through the use of cooked foods for thousands of years, and our old grandfather, the gorilla, only five feet in height, can encounter five or six of us at once owing to his superior strength and agility. In early times the apes were classified as men. They were ape-men: we are men-apes. There are tribes of hairy men at the present day—men with long hair attached to their whole face, as well as covering their entire bodies. The human embryo cannot be distinguished from that of the anthropoid ape until just before birth, and even then the distinction is very slight. Some of the lower orders of monkeys have a slight inclination to flesh-eating, but only when their natural food becomes scarce; and this tendency is quite in harmony with the observation, quoted by Professor Huxley, that "a starving sheep is as much a carnivore as a lion." The guiding principle which concerns us here is, that animals which consume their natural food, namely, that which is most suited for their structures, have the best chances to survive in the struggles for existence, other conditions remaining the same. If we can now ascertain what the natural food of the ape is, we have solved the problem, so far as man's natural diet is concerned; but we must also bear in mind that man has not evolved in the slightest degree in the direction of carnivorism, for he has not adopted nature's methods for such a process of evolution, and the only conclusion is, that flesh-eating is purely a habit which, like all our other depraved habits, can be gradually broken off. The question is not whether we, with our weakened vital organs, can return to our natural diet, but whether we suffer less injury from the adoption of our natural foods than from persistence in our artificial methods of living. If we cannot retrace our steps, civilized man is doomed to extinction; otherwise, no matter how slowly we ameliorate our dietetic habits, there is hope for the perpetuity of civilisation, providing other problems connected with our being and development can also be placed on a scientific basis. Many instances may be cited in which tribes have changed their dietetic habits, often very suddenly, without a marked depreciation in their development. The prince of meat-eaters, the Eskimo, when his food becomes scarce, has been known to browse on trees, and to live on and store up for future consumption leaves, bark, roots, &c., which they eat with great avidity. The California Indians, when deprived of their hunting-grounds through the advances of civilisation, have been known to feed on the juicy heads of wild clover, and have relished nuts and berries. Jenkinson relates that in Russia, three centuries ago, the inhabitants of Moscow supported life in summer by eating roots and grasses, and in winter by bread made from straw; and the bark of trees was relished "as good meat with them at all times."

The natural food of the monkey-tribe is grain, nuts, bread-fruit, apples, and bananas. The fruit-list, however, may be widely extended.

We are here, unfortunately, confronted with two schools of vegetarians. The one takes chemistry as the basis of dietetics, while the other accepts genealogy and biology as the basis, making chemistry a subservient science. Another mark of distinction is, that the chemical school inclines to take man as he is, while the biological school strives to make

man as he ought to be. The weakness of the chemical
school is that, under the treatment of its disciples, the
patients are growing worse and worse, no odds what school
of medical quackery they belong to. The same forces which
have made us bad continue to make us worse ; yet this is
the orthodox school at the present day, and cannot be sharply
separated from the defenders of the mixed diet. Its theories
cannot be defended with greater success than those of the
orthodox school of political economy ; and the same disaster
awaits the disciples of chemistry. Seeing that civilised man
eats cooked foods, they rush to the conclusion that man is a
cooking animal, and they prescribe his rations in sympathy
with this theory. They entertain, moreover, the theory that
man, by virtue of the skill he has acquired in the art of
cooking, milling, spicing, drugging, &c., can, with impunity,
prepare almost all products of the vegetable kingdom as
articles of diet ; and that, although meat-eating is objectionable, the products of the cow and the hen are suitable articles
in our bill of fare. On the other hand, the disciples of the
biological school contend that man can no more evolve into a
cooking animal than he can into a carnivore ; that his natural
foods possess the most delicious and delicate flavours when
eaten raw, which is another proof of our Simian origin ; that
the nutrients in our natural foods have the proper relation ;
and that cooking is the process of destroying the unpalatable
flavours of our unnatural foods, and has not been instituted
for improving the flavours of our natural foods. Finding that
chemistry proves the doctrines of the biological school to
be true, while biology cannot be made to harmonise with
chemistry, I am obliged to accept the former as being the
scientific school, and its disciples are therefore the only true
vegetarians. The disciples of the chemical school I am obliged
to classify as economic vegetarians, mainly for the reason that
they are dealing with economic men, and their establishments
are instituted for the purpose of drawing crowds and making
money. There are about forty of such so-called " vegetarian
restaurants " in London ; and their standpoint for the rejection of meat is based mainly upon " moral " grounds. However, they have lost all title to the name of moral vegetarians,

for the products of the cow and the hen are found in their bills of fare. It is immoral, they contend, to slaughter innocent animals; but the superannuated cow or hen suffers the same fate as the beef steer. Indeed, it is more immoral, one would think, to butcher the poor cow who has devoted her life to the gratification of our wants than to slaughter the bullock who has led a life of voluptuous indulgence. It makes no difference to the cow whether she is slaughtered for human consumption or as food for dogs or worms. If our "moral vegetarians" established a hospital for the support of cows and hens which have survived their usefulness, they could then consistently parade their moral assumptions. To their credit, however, be it said that they, like the scientific vegetarians, reject alcoholic beverages, although, differing from the scientific vegetarians, they still retain the alkaloids (tea, coffee, cocoa).

Fortunately, however, so far as this inquiry is concerned, there is a method of uniting these two schools; for we shall see that each requires substantially the same area of land for its support, providing dairy and poultry products be rejected. Indeed, these products must be rejected from a vegetarian inquiry. My examination now is to inquire into the area of land required to support (1) the vegetarian, (2) the meat eater, and (3) the mixed dieter. In order to make these calculations two data are necessary: (1) the productiveness of the land, and (2) the average daily ration to be consumed. In order to avoid the details required to satisfy the orthodox school of vegetarianism, I shall accept their ration, although it is higher than is warrantable in the experience of the scientific vegetarian, thus evading captious criticism. In addition to what has already been said against economic vegetarianism, I shall merely add here that the theory of Liebig, to the effect that muscular power was expended at the expense of muscular tissue, has long since been exploded by an array of practical experiments, it now being known that muscular exertion is produced by the expenditure of the carbonaceous compounds of the ration consumed. This theory is at the bottom of the mischief relating to animal diet; for, animal foods being highly nitrogenous, it was supposed that they best sustained physical exertion. This theory is also weak from the standpoint that there are vegetable foods, such as leguminous seeds, which contain at least as large a percentage of nitrogen, which enters so largely into the composition of muscular tissue, as does a large percentage of meat from well-fed animals. But, like economic theories in the minds of our economists, the nitrogen idea has got into the heads of our learned doctors, and it almost requires a surgical operation to extract it.

The most numerous and exhaustive experiments in dietetics have been conducted in Germany, and as the German standards do not differ essentially from those laid down by our own authorities, there is no liability for the admission of error. The German standard for a middle-aged adult of average weight (about 150 lbs.) and taking moderate exercise is a daily ration containing 4.3 oz. of protein, 2.1 oz. of fat, and 18 oz. of carbo-hydrates[1]. This standard gives an

1 In feeding experiments with cattle, which I shall have to discuss presently, the word *albuminoids* is used instead of protein or protids, but in order to assimilate the practices, I shall use the word *protein* all through. In human dietetics, the word amyloids is sometimes used instead of carbo-hydrates. In deference to universal practice, however, I shall continue to use the expression *albuminoid ratio*, the expression protein ratio not being used. Protein comprises a variety of nitrogenous compounds, and as their main function is to produce muscular tissue, they are sometimes called flesh-formers, while the fats and carbo-hydrates, being essentially the source of fat and heat, are sometimes called fat or heat producers, and they are under ordinary circumstances almost exclusively the source of muscular power. The carbo-hydrates embrace sugar and starch, with the small percentage of cellulose dextrine, &c., found in plants, and all these nutrients have one characteristic in common, namely, that they are composed of carbon and the elements of water. Fat and carbo-hydrates may be grouped together under the name of carbonaceous or non-nitrogenous compounds, the main nutritive element being carbon, but fat does not, like the carbo-hydrates, contain its oxygen and hydrogen in the same proportion as it is found in water (H_2O). Protein contains nitrogen, as well as carbon, hydrogen, and oxygen, and the nitrogenous compounds have different names in different foods—such as *gluten* in flour, *albumen* in the white of an egg, the *fibrin* in the blood, the *myosin*, *snytonin* of flesh (muscle), *casein* in milk, &c. The protein may be converted into its equivalent of nitrogen by dividing by 6.25, or by multiplying the nitrogen by 6.25 the quantity of protein may be determined—in other words, 16 per cent. of the protein is nitrogen. It is not of much consequence to pay attention to the relation between the fat and carbo-hydrates in any given ration, for the one may, to a very large extent at least, be substituted for the other. Indeed, some authorities reduce all rations to their equivalents of nitrogen and carbon. This transformation is easily made when it is known, in addition to the above figures, that fat contains about

albuminoid ratio of 1 : 5.4. This is the chemical standard, but the chemist has no means of ascertaining whether the ration should be drawn from the animal or the vegetable kingdom, much less from any species of plant or animal. This is a problem for the biologist; but I shall show that this orthodox ration may be obtained from plants as well as animals. In order to settle this question, let us consider the following table, being the average of a large number of analyses :—

Chemical Composition of Foods.

Name of Foods.	Water.	Protein.	Fat.	Carbo-hydrates.	Ash.	Albuminoid Ratio.
Wheat .	14.0	12.4	1.8	70	1.8	1 : 6
Oatmeal .	14.8	12.6	5.6	64	3.0	1 : 6.1
Fruit (fresh)	84.0	0.5	...	10	0.5	1 : 22
Nuts . .	4–6	16–24	55–65	12–14	2–3	1 : 6–10
Legumes .	12–14	22–24	1.6–1.8	49–53	2.5–3.2	1 : 2.2–2.6
Beef . .	55–77	11–21	2–25	...	0.7–1.5	1 : 0.2–5.5

Based upon the above analyses, a complete ration can be formed from any one of the foods named in the table, fruits and legumes excepted, fruits having too wide an albuminoid ratio, and legumes too narrow. With reference to beef, however, it must be very fat in order to make a complete ration by itself, and as to wheat and oatmeal, the average ratio (1 : 6) is sufficiently near the standard for all practical purposes. Some varieties are narrower than the standard, namely, about 1 : 5, and although some varieties of nuts give a ratio of 1 : 10, yet there are others, notably almonds, which analyse 1 : 5 to 1 : 5.5. Although in general it may be said that meat is the most nitrogenous food, yet the table shows that a very small percentage of legumes—peas, beans, or lentils—in a ration of the more carbonaceous foods would

76.5 per cent. of carbon, starch 44.4 per cent., and sugar 40 per cent. Protein contains 52.7 – 54.5 per cent. of carbon, and the percentage of nitrogen may range between 15.5–16.5 per cent. *Albuminoid ratio* is the relation between the quantity of protein and carbo-hydrates, the fat being first multiplied by 2.4 and then added to the carbo-hydrates, which gives the carbo-hydrate equivalent of fat.

give the proper ratio. Even with vegetables, potatoes, and rice, containing a ratio of 1 : 8 to 1 : 10, the ration can be sufficiently narrowed down by the use of legumes, animal foods therefore being quite unnecessary. Another item, which must not be overlooked, is the richness of oatmeal, fruits, nuts, and legumes in their quantity of salts, wheat and beef standing very low in this important constituent of our foods. Nuts are nutritively the most valuable of all the foods named in the table, one pound shelled being equal to three or four pounds of boneless meat. Fruits, although deficient in dry solids, are very valuable on account of their succulence, their organic acids, their organised water, and their saline constituents.

My remarks, however, are based upon the theories of the chemical school which yet sways our dietetics, including economic vegetarianism, and I have shown that, even from the chemical standpoint, animal foods play no part in the standard ration. A weakness of this school is, that its adherents fail to take into account the injurious effects of cooking our foods, which we shall presently consider. Another weakness in economic dietetics is, that the consumers narrow the ratio of their rations by eating large quantities of animal foods, and then widen it by the consumption of large quantities of sugar, fat, butter, and other carbonaceous compounds, which have no value whatever in a scientific ration. Although protein is regarded as being so very valuable, yet "rich foods" are those which are largely composed of fatty substances, and it is unquestionable that the average ration of the economic man has a ratio much wider than the standard. Taking the science of biology as the basis of dietetics, the albuminoid ratio given by the chemists is much too narrow, especially for consumers who take plenty exercise and pay due attention to hygiene matters generally. The scientific vegetarian requires neither legumes nor meat in his rations, and as many tribes thrive on fruits alone, which practice has also been successfully tested by scientific vegetarians, the chemical standard falls to the ground. The fact that so many people live almost exclusively on potatoes and rice, which have an albuminoid ratio of 1 : 10, although these foods are not suited for human consumption, is another convincing proof of the fallacy of accepting chemistry as the basis of dietetics. Scientific vegetarians have obtained excellent results by rations having a ratio of 1 : 8 or thereabouts; and an objection to a ration exclusively composed of fruits is, that the quantity to be consumed is too bulky for a large majority of them. Another objection is, that fruits are very digestible, and do not therefore afford sufficient exercise to the digestive organs. It must not be inferred from these observations that the scientist takes the carbonaceous compounds as the basis of the ration, for on this point he agrees with the orthodox doctors, namely, in accepting the protein as the basis. Life can be sustained on protein alone, including the necessary quantity of saline constituents, for it contains over 50 per cent. of carbon, so that the carbonaceous compounds may be regarded essentially as economisers of the protein. All rations must contain a fairly constant supply of protein, but the non-nitrogenous compounds may vary considerably, depending, (1) upon the amount of physical exertion expended, (2) whether these compounds are composed essentially of fat, sugar, or starch, and (3) the temperature of the medium in contact with the skin, i.e., the quantity of clothing worn, or the temperature of the air. The cooler the skin, the greater is the draught upon the carbonaceous compounds contained in our rations. If these compounds are composed of sugar or alcohol, they are consumed much more rapidly than when composed of fats or starch, and as fruits contain large quantities of water and sugar, they form the beverage of the scientific vegetarian. On the other hand, since meat contains no sugar or starch, alcohol seems to be the favourite fruit of the economic man, especially if he does not get a sufficiency of artificially prepared sugar; and as to the starch, a great deal of it is lost through the process of cooking, the food then being so sloppy that it cannot be efficiently masticated.

Adopting the orthodox standard already mentioned, viz., a daily ration containing 4.3 oz. of protein, 2.1 oz. of fat, and 18 oz. of carbo-hydrates—and for this purpose it is only necessary to consider the protein, for any ration containing an ample percentage of this nutrient will not be lacking

F

in fat or carbo-hydrates—the following ration may be obtained :—

```
 2 oz. nuts (shelled weight) containing 19½  per cent. protein 0.39 oz.
32  „   (2 lbs.) fruit              „       0.5    „      „    0.16 „
30  „   (1.87½ lbs.) grain          „      12.5    „      „    3.75 „
───                                                          ─────
64 oz.   = 4 lbs.              Total protein in ration  .  .  4.3 oz.[1]
```

Before proceeding to inquire into the area of land required to produce this ration, let us consider the question from a purely practical standpoint. The total weight of food in the above ration is 4 lbs. By comparing this with the quantity of food given to the soldiers in different countries, we get the following results: The daily ration for the British soldier, which includes 12 oz. of meat, is 4 lbs. 4 oz. ; the French soldier (with 10 oz. 9 drs. of meat), 3 lbs. 3 oz. 12¾ drs. ; the German soldier (peace ration), 2 lbs. 12 oz. 4 drs. ; the American soldier (with 1½ lb. of meat), 2 lbs. 10 oz. 10¾ drs. ; the Russian soldier (with 3 oz. of meat), 2 lbs. 10 oz. 3½ drs. The average ration for all these countries is 3 lbs. 2 oz., the proof being that my ration, which very nearly equals that of the British soldier, cannot be objected to on practical grounds.

[1] The nuts usually consumed are almonds, which can be purchased ready shelled in London for 7d. to 14d. per lb. Two lbs. of fruits may be considered a large bulk for some people, but in practice a portion is usually eaten dried, especially figs, dates, sultanas, and currants—that is, the 80 to 85 per cent. of water is reduced to 18 to 25 per cent., so that 1 lb. dried has about the same nutritive value as 2 to 3 lbs. fresh. These fruits can be purchased in the London markets for 4d. to 6d. per lb. It is more scientific, however, to eat fresh fruits, and when apples, &c., become scarce, fresh fruits can be purchased in tins for 4d. to 6d. per lb. The cost of the scientific ration is 6d. to 8d. per day, but may range between 4d. and 1s. according to the season and other circumstances. Sometimes, for the sake of convenience, the grain, nuts, and dried fruits are ground and formed into loaves or cakes, but not cooked, the mass being held together by the moisture in the fruit ; but, apart from the labour, there are two objections to this process : 1. No moisture in any form can be added without creating an unnatural thirst while eating, and dietetic perfection cannot be attained until nothing is drunk except fruits. 2. Any form of tampering with our natural foods (grinding probably excepted) destroys their delicious and delicate flavours, which can only be preserved by eating each food separately. Grinding should be dispensed with as much as possible, although people with weak or imperfect teeth or jaws can thrive better on finely ground raw meal than when cooked, for the pressure of the gums causes insalivation, which is the most essential process. Scientific rations are now used as a remedy for a large number of ailments.

From the practical experience of scientific vegetarians, however, it is found that, in addition to the quantity of nuts and fruits mentioned, they cannot consume on an average more than 1 to $1\frac{1}{4}$ lb. of grain, while the ration contains $1\frac{7}{8}$ lbs. There is no way of accounting for this difference except by the enormous loss in the process of cooking. This fact has also been proved by all accurately conducted experiments with domestic animals. This loss is due to the coagulation of the protein by cooking, and this explains the reason why the economic ration must have a narrower ratio than the scientific, the coagulated protein being less digestible; but the loss of nutritive value due to insufficient mastication and insalivation must also be taken into account. These soldiers' rations are, besides the quantities of meat named, composed mainly of bread and vegetables; the tea, coffee, and salt, having no nutritive value, must be classed as luxuries.

No objection can be raised against my adoption of the scientific ration in preference to an economic one; for, as we shall see, essentially the same area of land is required in both cases, although I have given the critic an advantage in assuming the quantity of food consumed in both rations to be the same, despite the fact that vegetables form the basis of the economic ration, and fruit that of the scientific. The quantity of nuts is somewhat arbitrarily chosen, having the basis of economy rather than science, for where nuts are abundant, more than 2 oz. may be added to the daily ration. No strict scientific rule can be given in this particular.

There is another difficulty with which I have to contend, namely, that our statistics are prepared for economists, not for scientists. Money values have nothing whatever to do with scientific inquiries; and although our agricultural abstract contains the average yields per acre for grain, I have had to depend upon personal inquiries for the yields of fruits and nuts. My calculations have been based upon estimates received from fruit-growers in the vicinity of London, and I have received my estimates for nuts from farmers in Kent county who grow "cob" nuts for the London market. Unfortunately, however, I have no analysis of the "cob," but as the variations in a number of varieties are limited between

16 and 24 per cent. of protein, I cannot be far astray in assuming the "cob" to contain 19½ per cent. Any objection raised against this analysis will be more than counterbalanced by the fact that nuts grown in forests and on large trees are more nutritive and productive than those grown under cultivation.

It now remains to connect this ration with its equivalent in land acreage. Of the two systems of apple-culture, dwarf and standard, the average yield per acre in the mixed system is 7½ tons, the standard being somewhat more and the dwarf somewhat less. Taking apples as the basis of our fruit-culture, this estimate gives us an annual yield of 16,800 lbs. per acre; but this is not the total yield of fruit, for it is a general practice to grow strawberries, tomatoes, mushrooms, &c., between the rows of trees, and the larger varieties of small fruits are also sometimes grown in the orchard. I have estimates of tomatoes yielding an average of 7 tons per acre, and of strawberries yielding four times the quantity of apples. Assuming one-fourth of the space in the orchard to be devoted to strawberries, the total yield of fruits will be represented by double the yield of apples, or 33,600 lbs. The yield of potatoes for the United Kingdom is given in the agricultural abstract, the average for the past five years (1884–1888) being 4.66 tons, which, compared with the yield of tomatoes above given (7 tons per acre), amounts to almost exactly the same thing, for 4.66 tons containing 1.8 per cent. of protein = 0.084 ton, and 7 tons tomatoes containing 1.25 per cent. protein = 0.087 ton. It is important to bear these figures in mind, for they are connected with the figures given in the ration, to prove that the area of land is substantially the same, no odds whether scientific or economic vegetarianism be adopted. The yield of "cob" nuts, namely, 1 ton per acre—and I have ascertained that almost exactly two-thirds of the weight is shell—is now represented by 750 lbs. of edible portion. According to our agricultural abstract, the average yield of wheat for the past five years (1884–1888) is 29.6 bushels per acre; oats, 37.22; beans, 24.93; peas, 23.87. These grains may now be grouped into two classes, the wheat, peas, and beans, yielding an

average of 26 bushels per acre, being essentially the grain basis
of the economic vegetarian. Wheat is a very artificial product, and is therefore almost destitute of flavour; when eaten
raw it is apt to gum in the mouth, the most nutritive varieties
are apt to be flinty, as well as gummy, and when ground it is
nasty stuff to handle. For the scientific vegetarian, the basis
of the grain diet is oatmeal, even barley and rye being preferable to some varieties of wheat, and it has the most delicious
flavour, as well as being the most nutricious of all the cereals.
Oats contain about 55 per cent. of well-dried oatmeal, which
equals a yield of 20½ bushels of oatmeal per acre; but, calculated with the same percentage of moisture as wheat, the
yield would be approximately 25 bushels per acre, weight
about 50 lbs. to the bushel. This classification gives a slight
advantage to the economic vegetarian, but the difference is so
slight compared with the concessions I have made that no
criticism need be anticipated. Making due allowance for seed,
it will be fair to classify these crops as grain yielding 23 bushels
per acre, the average for the four grains being 25.8 bushels.
Connecting these estimates in tabular form, we obtain the
following results:—

```
1 acre fruit yields 33,600 lbs. containing 0.5 per cent. protein .   . 168 lbs.
1   ,,   grain  ,,   1,380  ,,        ,,       12.5    ,,         ,,   . . 172  ,,
1   ,,   nuts   ,,     750  ,,        ,,       19.5    ,,         ,,   . . 146  ,,
                                                                        ________
                          Total protein per 3 acres  .   .   .  . 486 lbs.
                             ,,      ,,     ,,  1 acre  .   .   .  . 162  ,,
```

Comparing these figures with one another, and also with
tomatoes and potatoes, let us see who has the advantage, so
far as area of land is concerned, the economic or the scientific
vegetarian. Tomatoes yield 194 lbs. of protein per acre; but
as this vegetable is common to both schools, the scientific
vegetarian consuming large quantities of this and all other
vegetables which may be eaten raw, such as melons and
cucumbers, all criticism here must be abandoned. Potatoes
yield 188 lbs. of protein per acre, but after deducting, say,
half a ton for seed from the total yield, the crop is reduced to
4.16 tons instead of 4.66, which reduces the protein to 168
lbs. per acre. It need hardly be said that the potato is

unceremoniously dismissed from the table of the scientific vegetarian. Nuts appear to stand the lowest on the list, but as the quantity eaten is small, as it is difficult to get accurate estimates even over so small an area, and as I am convinced that the above estimate is below rather than above the mark, the figures must be criticised with caution. Unfortunately, I have been unable to get reliable returns with reference to the yields of other vegetables and fruits. The unanimous conclusion must be that both vegetarian schools draw on substantially the same area of land for their support.

Having ascertained the yield of protein per acre to be 162 lbs.—2592 oz.—and as each individual requires 4.3 oz. daily, then 2592 ÷ 4.3 = 602.8, so that an acre will support one adult for 602.8 days, or 1.65 persons for a year; therefore 0.6 acre will be required to grow sufficient food for the yearly support of one middle-aged healthy person of average weight and taking appropriate exercise. Taking the United Kingdom as our model, the population (about 300 per square mile, which is greater, with one or two exceptions, than any other country in the world), being about 37,100,000, we can now easily calculate how far our cultivated area will go towards the support of our own people. But this population includes children as well as adults, also aged persons, who do not consume so much as the standard ration. According to our last census returns, about 35 per cent. of our population is under fifteen years of age. In order to evade the lengthy calculations required to work out rations for people of different ages, I shall assume—which assumption is another point yielded in favour of the critic—that the average youth between the ages of birth and fifteen years is equal in consuming capacity to half an adult. Should any doubt arise, it will be fully subjugated by the percentage of persons over sixty years of age, who consume less than middle-aged people. From these data our present population, reduced to adults, equals roundly 30,000,000, and as each requires 0.6 acre for his support, our total population will require 18,000,000 acres. These figures are now to decide whether our isles are over-populated, so far as the production and consumption of food are cosnidered.

Of the total land area of the United Kingdom, namely,

77,799,793 acres,[1] 47,876,814 acres are under cultivation, exclusive of "heath and mountain pasture land, and of wood and plantations," by which it will be seen at a glance that our present population could be supported on a little over one-third of our cultivated area, almost exactly three-eighths of such area, or $37\frac{1}{2}$ per cent. If we now take the difference between our total and cultivated areas, which is very nearly 30,000,000 acres, of which, it is said, 12,000,000 could be profitably cultivated, and a larger percentage of the remainder could certainly be devoted to fruit and nut-bearing trees, it will be no extravagance to assert that our present population could be supported from our waste areas.

Having looked upon this picture, let us now look upon that. After cultivating three times the area of land required for our subsistence, we then import annually ten million bushels of fruit, over 140 million cwt. of grain (including their equivalent in flour or meal), nearly two million head of domestic animals, about nine million cwt. of different kinds of meat (including fish), five to six million cwt. of butter, margarine, cheese, and eggs, about $6\frac{1}{2}$ million cwt. of sugar, besides various other articles too numerous to be mentioned. We import more grain alone (including its mill products) than is sufficient to support our whole population, and yet we are so over-populated that we must banish annually from our shores 250,000 of our brothers and sisters. Economicism, moreover, has been instituted for the purpose of making nations populous, and consequently also wealthy. Here we have reduced to actual figures the value of the theory that wealth must be based on scarcity—that abstraction and negation are the most substantial foundation of economic science.

That we have here struck a violent dualism, namely, science *versus* economicism, must have already been observed by the critical reader. Science converts millions of acres of our fertile areas into a worthless waste, there being insufficient population to consume its products, and thus bankrupts rent and trade. The scientist is driven from every standpoint to seek a form of wealth which is based upon abundance, not upon scarcity, and as land is now abundant, another problem

---

[1] Agricultural Abstract, 1888.

to be solved is how to make the products of the land plentiful. The labourer may here be driven to despair lest I deprive him of work which he so eagerly seeks, and he should be fore-warned to know that this is precisely what I intend to do. Labour is sought for the purpose of obtaining the means of subsistence, and if the labourer can obtain the objects of his desire without labour, he will be the last person to complain. The theory that people have a right to labour is one of the wildest of all abstractions. Putting it in the mildest form, the man who claims the right to labour should be able to show the logical effects of his exertions. By virtue of the nature of my inquiry, I can permit the existence of no form of labour which tends to the destruction of the labourer or any of his fellows. The secret of all the poverty, all the crime, which exists in the civilised world, rests on the theories that wealth is based on scarcity and that people can claim the right to labour.

However, before we can nicely determine whether land in the United Kingdom is plentiful or scarce, we must also ascertain the area required to support our population in clothing, beverages, &c. This inquiry belongs to another chapter. The theory, tacitly or openly expressed, that our clothing, like our beverages, is intended to keep us warm in winter, cool in summer, and to have a neutral effect in spring and autumn, is of no value to the scientist until the economist informs him whether the heat should be supplied by the combustion of fuel (food) within the body, or whether the same object can be attained by external protection from cold. If the economist further urges that man has developed into a shame-animal, thus requiring clothing to hide his shame, the scientist still wants to know whether shame is an abstract idea, or whether it belongs to the "innate moral sense"—also whether man should not hide his head for shame rather than his body. Until these questions be decided, there is no sense in wrangling over the interior dualism, namely, woollen *versus* cotton clothing. It is not my place to deny that all our paraphernalia are adjusted for the good of trade, but it must not be forgotten that my inquiry is adjusted for the good of humanity.

It is in order here to say a word about cooking. I have

pointed out the fact that we require more land to support us on cooked foods than upon the same foods eaten in their raw state ; but from this standpoint alone, I do not attach much importance to the question. One of the most essential branches of my inquiry is the emancipation of labour, and especially the emancipation of women. There is a superstition afloat that a great deal of our misery is caused by bad cooking, and that the only remedy is the establishment of institutions for acquiring the art or science of cookery. Before the scientist consents to be taxed for furnishing respectable employment in this particular, he requires an answer to the following questions: 1. Should we produce foods whose flavours are adjusted to our natural tastes, or accept the productions of nature as they are, adjusting them to our tastes by destroying their natural flavours and making them palatable by cooking, salting, spicing, &c. ? 2. What is the effect of cooking upon each of the nutrients contained in our food ? 3. If cooking is intended to ease our masticating and digestive apparatuses, why do our brains and muscles require exercise and our digestive organs none ? 4. If our food is to be eaten hot, can (a) heat in the body be produced naturally in this manner, and (b) what is the sense in thus producing heat under summer temperatures ? 5. If the object be to furnish employment, then (a) can other employment not be obtained, and (b) what is the tendency of that form of employment by which people are obliged to stand over a hot cooking-stove, thus producing heat in the body by external application in addition to that produced within the body by eating hot foods ? These questions will be more thoroughly answered in subsequent chapters ; but meanwhile I shall quote the words of an eminent authority[1] who speaks from long experience and observation :—

"Of all the artificial forms of treatment to which foods are subjected, that of cooking is the most universal, and therefore demands our special attention. If we rightly consider the influence of this process upon all the natural properties of a plant, we must concede that it is in almost every case injurious, and that it should be

---

[1] *Bread and Fruit: A Scientific Diet.* By Gustav Schlickeysen ; translated from the German by Dr. Holbrook.

dispensed with, so far as our present habits of life will admit of, and with a view to its final and complete disuse. The natural fluids of the plant are, in great part, lost in cooking, and with them the natural aroma so agreeable to the senses and so stimulating to the appetite. The water, supplied artificially, does not possess the same properties as that which has been lost, and all the less so since it has been boiled. The cellular tissue of the plant loses all its vitality, and ripe uncooked fruits and grains, with their unbroken cellular tissue, their stimulating properties, their great content of water, sugar, and acids, and their electric vitality, are calculated to impart to the human body a rosy freshness, to the skin a beautiful transparency, and to the whole muscular system the highest vigour and elasticity. Uncooked fruits, especially, excite the mind to its highest activity. After eating them, we experience an inclination to vigorous exercise and also an increased capacity for study and all mental work ; while cooked food causes a feeling of satiety and sluggishness. Not only do plants lose their vital, but, to some extent, also their nutritive, properties when cooked. The vegetable acids and oils, the latter being of especial value in the development of the bony structure of the body, are, by cooking, dissipated, while the albuminoids are coagulated, and thereby less easily digested, so that the nutritive value of the food is reduced to a minimum. Another injury that results from cooked food is that caused by the artificial heat. All heat excites, through expansion, an increased activity, but this activity is not normal in the case of food eaten hot. Again, the sensory nerves of the lips and the nerves of taste are weakened by hot food to such an extent that they no longer serve as an infallible test of its quality, and hence articles that seem in the mouth to be palatable and good may be very injurious to the system both on account of their natural properties and their artificial heat. In a similar manner, the sense of smell is blunted ; and not less injuriously does hot food act upon the teeth, the enamel of which is destroyed, rendering them unfit for their work of mastication, in consequence of which the food passes unprepared into the stomach. The eyes are also injured by the action of hot food upon the nerves connected with them. That condition of weak and watery eyes, so apparent in the habitual drunkard, exists in a certain degree with all whose systems are enervated by hot and stimulating foods. But the greatest harm from hot food is caused in the stomach itself, the coats of which are irritated, reddened, so that they lose their vigorous activity and capacity for the complete performance of their natural functions. The blood, excited by the heat, flows in excess to the stomach, and thence feverishly through the body. One result of this is the flushed condition of the head after eating. Hot food also causes excess in eating, so that it is rather by a sense of fulness and oppression than by a natural satisfaction of the appetite that one is prompted to cease eating. An evidence of the weakness of the stomach by hot food is seen when one eats an apple immediately after the usual hot meal. Fruit thus taken lies like a stone upon the stomach, the enfeebled nerves being injuriously affected by its presence ; whereas, in their normal condition, they are stimulated to a most agreeable activity by it. From the abuse of the organs of digestion result a host of diseases. A life-long weakness of the gastric nerves, with cramp and inflammation of the stomach, are the common fruits. To this cause also is attributable the almost universal prevalence of colds, which are the direct result of unnatural temperature conditions of the body. The blood, artificially heated, causes an excessive perspiration, since it produces an increased, but injurious, activity of the skin ; and upon the least change of temperature, the perspiration is condensed upon the body, and causes colds and stiffness, and this all the more certainly when the blood is impure and the tissues overloaded. From the same prolific cause results all the uneasiness and languor experienced after eating hot food. But the evil effect cannot be overcome by the usual after-dinner nap. This cannot replace the elements lost from the food, or give the enlivening impulse experienced after partaking of ripe fruits in their natural state. It is, indeed, argued that our northern climate requires that food should be eaten hot as one means of maintaining the bodily temperature ; but if this be true of man, it must also apply with equal force to all animals, and since man alone seems to require hot food, the argument loses its force. In the polar regions, the conditions of animal life show plainly that the natural process of generating heat is not by putting heated substances into the stomach, but by the normal action of the vital forces upon the food taken in its natural state. Greater thirst is experienced after eating cooked than uncooked food, and this results both from the change which the food has undergone and from the perspiration caused by the increased heat of the body. The artificial solution of the food impairs its nutritive properties, and weakens the natural functions of the body by depriving them of their natural employment ; and this has been so long continued that we are now almost incapable of digesting uncooked grains, so that their enlivening and invigorating action is almost unknown. The modern kitchen has thus perverted the natural appetite and enfeebled the natural powers. It has also fostered injurious customs, and introduced articles of diet that would otherwise have been excluded. Only through its aid can the flesh of animals be rendered palatable. Its abolition gradually, if not at once, would contribute much to restore man to his normal dietetic conditions, and would exclude the most injurious parts of his present diet."

The above extract is a pretty specimen of man's supernatural powers of objectivity—his boasted mastery over natural laws. The orthodox doctor holds a position identical with that of the economist, and is, moreover, a great aid to economicism ; for he induces the belief that a specific can be found for the violation of natural laws. When he finds out the error of his ways, and ceases to enjoy vested rights in the manufacture of patients, his profession will develop into a science, instead of degenerating into the art of quackery.

Unless he can show the advantages to be gained by man's attempted evolution into a cooking-animal, his profession must go the way of all other abstract theories.

The objection that our foods must be cooked because domestic plants and animals themselves, being the result of economic husbandry, are artificial productions, is the direct reverse of a scientific truth; for cooking, like artificial agriculture, has the effect of widening the albuminoid ratio, the former by coagulating the albuminoids, and the latter by producing an excess of carbonaceous compounds in agricultural plants and domestic animals. Plants and animals plucked from the field of nature are not only more nitrogenous in their composition, but also more healthy and more highly flavoured than the same species under domestication. The excessive carbon, indeed also some of the nitrogen, is not healthy tissue; and it is, therefore, a great mistake to suppose, as many have done who have rather suddenly adopted the scientific diet, that a loss of weight is necessarily accompanied by a loss of health or strength. Indeed, the reverse is very frequently the case.

Practically we are a toothless generation, and so long as we continue to cast upon our stomachs those burdens which belong to our jaws, there can be no solution of the social problem. Making slaves of our stomachs and lords of our jaws is the same principle which has caused the extinction of all civilisations, and our bodies cannot escape the same fate. The number of millers and cooks is a measure of the time during which civilised nations can exist; for a function cannot continue after the structure is gone. The moral conclusion must also follow, namely, the supplanting of natural by stupefactive pleasures—the generating of miserable weaklings, puny runts, who can look without compassion upon the sufferings of their fellow-creatures.

This chapter would not be complete without inquiring into the area of land required to sustain our population on a flesh or mixed diet. I shall therefore proceed, by the same method, to draw the ration already given from the animal kingdom, instead of the vegetable. In the science of cattle-feeding, the method is somewhat different in detail, although identical in principle. Albuminoid ratios are seldom spoken of in human dietetics, but, in my opinion, it is the most intelligible expression of relations, and is adopted by all experimenters in cattle-feeding. The most important difference in the two methods is the recognition only of the *digestible* portions of the nutrients in stock-feeding. The number of experiments have been so large that digestible coefficients have been determined, whereas in human dietetics, the number of experiments being less, and man being a more economic or artificial animal than the bullock or even the hog, determinations of digestible percentages have proved very unsatisfactory. Man, however, has his digestive coefficients largely under his control, based upon the quantity of mastication which the food receives, but this advantage has faint application when cooked foods are eaten. It is to be further noted that, in cattle-feeding, it is always understood that the rations are consumed in their raw state. If the foods were reckoned as being cooked, the animals would draw on a larger area of land for their support. It is not necessary to consider the different species of domestic animals separately, for the results will be found to be essentially the same whether we take the steer, the sheep, or the hog as the standard. As the most numerous and exhaustive experiments have been conducted with bovines, I shall select the beef steer as the basis of my calculations, and shall then compare the results with those of a milk diet. Chemistry has done little or nothing more than to prove that the natural foods of domestic animals possess the proper albuminoid ratio and other qualities, although a great deal of useful knowledge has been accumulated in the chemical laboratory with reference to the compounding of rations which are composed, mainly or entirely, of by-products or artificial foods. It should also be noted that the object in feeding stock is to find a ration which will produce the greatest weight in the shortest time, which requires a ratio somewhat narrower than under natural conditions, the animals receiving little or no exercise and being closed up in warm stalls. How this differs from human dietetics I am not prepared to say; but, judging from observation, I should think that the only aim of the economic man in this world is to eat and grow fat.

According to the highest authority on cattle-feeding,[1] whose results have been practically substantiated by our own great authority, Sir J. B. Lawes, a steer weighing 1000 lbs. when stalled requires, as an average of the three periods of fattening, 26 lbs. of dry solids in his daily ration, which must contain 2.7 lbs. of protein, 0.6 lb. of fat, and 14.8 lbs. of carbohydrates, making an albuminoid ratio of 1 : 5.5, these nutrients referring to the digestible percentages of the ration. For the purpose of compounding this ration, we require 13 lbs. of hay (clover mixed with timothy or other grasses), 5 lbs. of peas, 6 lbs. of oats, and 30 lbs. of roots. I have selected these foods because the yields per acre are given in our agricultural abstract, namely, hay 1.41 ton per acre from permanent pastures, turnips 10.14 tons, mangels 16.66 tons, peas 24.93 bushels, oats 37.22 bushels. As already stated, the peas and oats are averages for the past five years (1884–1888). The averages for roots are for the same period, but only the yields of hay for 1887 and 1888 are given in the abstract, of which the given number of tons is the average. I take the average of turnips and mangels, and call the yield 13.4 tons of roots per acre. One acre of each of the above crops will therefore give the following results :—

Hay	3,158.4 lbs. containing	6	per cent. digestible protein,	189.5 lbs.
Peas,	1,495.8 ,,	,,	20	,,	,,	,,	299.2 ,,
Oats,	1,265.5 ,,	,,	9	,,	,,	,,	113.9 ,,
Roots,	30,016.0 ,,	,,	1.2	,,	,,	,,	360.2 ,,
			Total protein from	4 acres,	962.8 lbs.
			,, ,, ,,	1 acre,	240.7 ,,

An acre thus produces 240.7 lbs. of digestible protein, and as the steer receives 2.7 lbs. daily, 89 animals can be supported from an acre for one day, or 0.24 animal for one year. In other words, it requires 4.16 acres to support a fattening steer for a year.

This figure may be assailed by some practical farmers who have good pastures carrying two bullocks or cows per acre, or who feed large quantities of ensilage, but the data are taken from the averages of their own returns. It is assumed that

---

[1] *Landwirthschaftliche Fütterungslehre*, by Emil Wolff.

the area of land drawn on is the same under winter and summer conditions, which is sufficiently correct for the purposes of this inquiry. Should any practical objection, however, be raised, it may be offset by the fact, as proved by the experience of farmers, that there is a disadvantage in raising steers from calfhood up to the stalling period, it being more profitable to purchase store animals. This is quite natural, for the land drawn on by the steer's dam for nine months before and about six months after calving is chargeable to the steer.

As already shown, 0.6 acre will support an adult human being, and as the steer requires 4.16 acres, or very nearly seven times more land, it is plain that if the steer's increase in weight will support seven men, it makes no difference, so far as the area of land is concerned, whether we live on vegetable or animal rations. What now is the consumable produce of this steer? This is obtained by ascertaining the average daily gain during the feeding period of one year. At the last annual cattle show of the Smithfield Club (1888), the average daily gain of the eighteen prize-winners from birth was 1.8 lb. per day, the average weight being 1833 lbs., and the average age 999 days ($2\frac{3}{4}$ years). But these animals were extravagantly pampered in competition and for choice markets, thus receiving larger rations than the standard, and are not therefore a fair criterion for our purpose. From all the data I can gather, a gain of 1.5 lbs. per day is a liberal allowance. For the first year, or thereabouts, the gain would be more; and afterwards, especially after the second year, it falls off considerably. Our steer, entering the stall at a weight of 1000 lbs., would be about seventeen months old, and it requires an animal of the best beefing breed to gain 1.5 lb. daily after that period. After being fed for one year, he would be nearly two and a half years old when slaughtered, which is about the usual age, despite the "baby-beef boom." Based upon this daily gain, the total gain for the year will be 547.5 lbs. live weight, which represents the produce from our 4.16 acres. The steer, now being classed as fat, will yield as carcass about 65 per cent. of its live weight, including fat on the kidneys and intestines, so that the area yields in butcher's meat 356 lbs., which makes a daily ration of 0.95 lb. for a meat-eater, uncooked weight; and as the carcass contains about 7 per cent. of bone, the edible portion is reduced to 0.88 lb. According to estimates made in the army, there is a loss of 25 per cent. in bone, gristle, and other uneatable substances, and a further loss of 25 per cent. in cooking, which would reduce the ration to about $\frac{1}{2}$ lb.; but as these estimates are from inferior carcasses, they are hardly a fair criterion for our purpose. Two-thirds of a pound would be a nearer estimate. Now, it is well known to every observer that the habitual meat-eater, who drinks a fair average of stimulating beverages and swallows the usual quantity of aperient drugs, will consume $\frac{3}{4}$ to $\frac{7}{8}$ lb. of meat daily in addition to the same quantity of vegetable food as that consumed by the vegetarian, who, as a rule, has no desire for alcoholic drinks. The stimulating properties of meat, in connection with its absence of carbo-hydrates, are at the bottom of our drinking habits, and the cooking of foods intensifies the evil. It is, therefore, no extravagant estimate when it is said that the meat-eater requires 4.16 acres more for his support than the vegetarian, even without reckoning the extra area which he imbibes. The Eskimo, when he can get no vegetable food, consumes 6 to 8 lbs. of flesh daily; but as this is mostly blubber and consumed in very cold regions, it is not a fair criterion for our conditions.

But these observations are merely practical. Let us now bring the question down to a scientific basis. The carcass being fat, it forms a complete ration, so that no addition of vegetable food is necessary, and the analysis will therefore be 55 to 60 per cent. of water (lean meat contains about 77 per cent.), 10 to 12 per cent. protein, 20 to 25 per cent. fat, 6 to 7 per cent. bone, and about 1 per cent. of ash in the muscle. Based upon the scientific standard already given, which contains 4.3 oz. of protein in the daily ration, 3 lbs. of meat per day will be required, raw weight, or 1095 lbs. yearly; that is, roundly, the produce of three steers; and as each steer requires 4.16 acres for his support, the man who lives on a meat diet will require three times this area, being at least 12 acres. In other words, he requires twenty times more land for his support than the vegetarian, and our present population would therefore require, in round numbers, 375,000,000 acres, or nearly eight times our cultivated area. The mixed dieter has now his choice between the 0.6 and the 12 acres. If our cultivated area were equally divided amongst our people, there would be 1.6 acre for each person, the population being reduced to adult equivalents, so that if he adopted a vegetarian diet, he would have exactly one acre over and above the area required to support him in food. This acre not being required for meat, and therefore also not for alcoholic beverages, the next problem to be solved is how far it will go towards his supply of clothing, which inquiry belongs to another chapter. He gets almost exactly one more acre of waste land, which should be sufficient to satisfy him so far as playgrounds, parks, &c., are concerned.

Before passing on to the discussion of the milk diet, we have already sufficient evidence to expose the fallacy of all attempts to solve the Malthusian and anti-Malthusian theories by abstract processes of reasoning. The question may now be asked whether the scientist is a Malthusian or an anti-Malthusian. He is neither; and therefore, also, not both combined. The Malthusians maintain that our isles are over-populated, and that emigration or restraint is the remedy; which is not scientifically correct, for our present population could be increased at least threefold without importing any food. That restraint is required no scientist will deny, but not restraint on population, but restraint on abstract theories and the terrible calamities which they have produced. Indeed, it would not be much of a figure of speech if I said restraint on education. On the other hand, the scientist is not an anti-Malthusian because he can devise no means for bringing manna down from heaven, and the population must therefore be restricted within the limits of the soil's productiveness at any given period. Agricultural science has proved that neither the soil itself nor the raw materials utilised for maintaining its fertility is an inexhaustible mine; although, by the exercise of scientific economy, the fertility, and consequently also the population, could be increased. Another problem connected herewith is the increase or decrease of vitality in our agricultural plants, which, however, belongs to a later chapter. All that is necessary to be understood here

is, that no matter how small the area of land may be which is
required to support an individual, there is some limit to that
area, and a corresponding limit to the population. The anti-
Malthusians are therefore unscientific when they espouse the
theory of Providential interference, or maintain that so long
as people are born with twice as many hands as mouths, there
can be no danger of over-population. This idea is based on
the theory that the productivity of the soil may be increased
in proportion to the quantity of labour applied. We shall see
that, in the long-run, the exact reverse is the case. From the
scientific standpoint, over-population takes place when the
products of the soil fall below the standard ration for each of
all the inhabitants of a given area, or all the world, according
to the condition of commerce. The scientist is, therefore,
obliged to devise methods for restricting population within
this limit without violating any natural laws. Whether the
scientist should be hanged for discouraging trade, or the economist hanged for discouraging science, may be elicited from
subsequent pages.

A word about the cow is now necessary for the comple-
tion of this chapter. All economic men, even including the
economic vegetarians of the orthodox school, are called upon
to defend the cow, for she has not one scientific point in her
favour, and the same arguments apply with equal force to the
hen. In speaking of rations composed of milk, there is a
prominent weakness which, I regret to say, makes our English
language poverty-stricken in all attempts to find adequate
expression, so I am obliged to use an Americanism, namely,
Thar ain't nothin' to chaw at, somehow. The beauty and pro-
fundity of this illustration cannot be fully realised until
we are able to comprehend how thoroughly befitting it is
when also applied to all abstract theories. There seems to
be a superstition afloat that our jaws perform their natural
functions while engaged in discussing abstract theories, and
that they would be overstrained if they also performed the
artificial function of masticating solid food. Milk is recommended on account of its ease of digestion, but this argument
is weak from the scientific standpoint, because our vital
organs require exercise as well as our other structures. If

more digestible foods are required, why do our economic authorities not recommend fruit? It is a strange inconsistency to scour the earth for digestible foods, and then put them through the process of milling or cooking in order to make them more indigestible, thereby encouraging the trade in pills, "digestive candies," and other nostrums. Like our brains and our muscles, our vital organs require both light and heavy exercise. The same nice principles of discrimination are exercised with reference to the flavouring of our foods. For example, we flavour our porridge with salt, and then flavour the salt with sugar. We eat hot foods and drink hot beverages in summer to sustain the bodily warmth, and then swallow ice-water and ice-cream to keep us cool. We build fires in our houses to keep us warm, and then open the windows to keep us fresh and cool. In short, so great are our powers of objectivity that we can carry any desired climate in our pockets, and can modify any product in the three kingdoms of nature to appease our hunger or tickle our nerves of taste. Let us connect these observations with the following remarks by Sir Henry Thompson, namely, "that a proportion amounting at least to more than one-half of the disease which embitters the middle and latter part of the life amongst the middle and upper classes of the population is due to avoidable errors in diet," and he proceeds to attribute the cause mainly to over-eating. The economic man is always eating and is never satisfied—always scrambling for wealth, wherewith he may enjoy still greater luxuries, and never gets enough. Meat, although not masticated by carnivorous animals to any appreciable extent, has the advantage of affording some exercise for our teeth and jaws, which cannot be asserted with reference to a milk diet, while the latter is equally injurious on account of its stimulating properties. We masticate meat because it requires no mastication or insalivation, and bolt our vegetable foods because they require to undergo these processes. Moreover, it is urged that mothers cannot raise their infants without the aid of the cow. This is the most suicidal argument in all economic theories; it is a complete relinquishment of the whole case. If our mothers had milk to spare for needy calves, there would be some hope for our race; and what is still worse, cows are getting economic so rapidly that many of them have not sufficient milk to raise their own offspring, leaving no surplus for the offspring of human beings. My inquiry will be a failure if 1 cannot find mothers who can raise their infants on their natural diet, and prevent human beings from being calves all their days. It stands to reason that if mothers do not themselves consume natural foods, they cannot produce natural food for their offspring.

. Before accurately determining the area of land required to support an individual on a milk diet, the scientist demands from the economist an answer to the following questions: At what age should a cow drop her first calf? What is the period of her milk-giving usefulness? What is to become of her carcass? Should the milk be consumed mostly fresh or sour, or converted into butter, cheese, condensed milk, or koumiss? What is to be done with the skim-milk and the whey? Should the milk be drunk cooked or raw? Should she be fed mostly in the stall or on the pasture, and what are her rations? To what breed does she belong? It is not the part of the scientist to solve these questions, for he must condemn the whole process; but, certain data being given, he can extend them to their conclusions—a very unpleasant task for the true scientist. The conclusions can only be here approximated, and the question must be disposed of briefly.

In order to utilise the same feeding ration as that consumed by the steer, as already considered, let us assume her weight to be about 1300 lbs., and let her give milk of average quality, namely, 87.6 per cent. water, 3.4 per cent. protein (casein and albumen), 3.4 per cent. fat, 4.8 per cent. milk sugar, and 0.8 per cent. mineral salts. As with the steer, she will require 4.16 acres of land for her support, without counting the 10 or 12 acres required to produce her before she drops her first calf at the age of $2\frac{1}{2}$ or 3 years. As a human adult requires 4.3 oz. of protein in his daily ration, he must consume 6.4 lbs. cf milk per day, or 2336 lbs. per year. Some cows do not give more than this quantity of milk yearly, but as our cow is large, well fed, and presumed to belong to a good milking breed, she will give three times this quantity, say 7000 lbs.—

that is, she will support three men. In other words, 4.16 acres will support three adults, or 1.4 acre will support one adult. This apparently shows better results than the beef steer, but proves the fallacy of the "three-acres-and-a-cow" theory supported by a certain wing of our politicians. If the three acres supported the cow, the cow would not support two adults. But the matter presents quite a different phase when the milk is converted into butter or cheese. Two ounces of butter per day is not a heavy consumption (45.6 lbs. yearly), and as the milk (except when the fat is separated by centrifugal force) gives about the same percentage of butter as butter fat (butter contains about 12 per cent. of water), 7000 lbs. of milk will represent 238 lbs. of butter; so that our 4.16 acres will be required to produce butter for 5.2 persons, or each person will require 0.8 acre to keep him in butter, or exactly one-third more land than is required to supply the vegetarian with all the food he consumes. From the scientific standpoint, butter is a perfectly useless and valueless article, for more wholesome and delicious forms of fat are contained in our natural foods, and should not be separated from them. As nobody can live on butter, the area of land required to support an individual on a butter diet cannot be determined, but from cheese we can form a complete ration. As milk produces 10 per cent. of cheese, the person who consumes $\frac{1}{4}$ lb. of cheese per day—and this is a small piece on a slice of bread—will require about the same area of land to keep him in cheese alone as the vegetarian requires for his complete rations. Cheese contains from 24 to 34 per cent. of protein, and it would therefore require about a pound per day to support a person on a cheese diet, which represents nearly $2\frac{1}{2}$ acres of land. These figures would be still more alarming if we calculated the area of land wasted in producing the cow.

Driving these estimates to their logical conclusions, the reason why we require so much land for our support is because we are too lazy to work our jaws—except for the discussion of abstract theories. The butcher, the miller, the cow, and the cook are the measure of our vital degeneracy. The poor people and the hogs, by virtue of some providential arrangement, I suppose, get the most nutritive portions of our foods, while the luxurious, or comparatively worthless, nutrients go to the rich. These luxuries are scarce because they require a great deal of land and labour for their production, and they are luxuries because they are scarce. The rich get the butter, the more worthless portion of the milk, and the poor people and the hogs get the protein—the most valuable nutrient. So also with the wheat; the hogs get the protein, and the men the carbo-hydrates. Even from our butchers' shops the gluttons carry away the fat, because it requires so much land and labour to produce fat animals, and the poor men get the protein, because the protein-producing breeds of domestic animals have no pedigrees, and are allowed to shift for themselves. It does not now require a strong logical faculty to see that the farther we remove ourselves from the laws of our being and development the more land and labour we require for our support. In other words, with the advance of economicism we draw on correspondingly more land and labour for our maintenance; and the same rule applies with equal force to our clothing, beverages, medicines, &c., and even to the area of land required to furnish us with light while penetrating the depths of abstract theories.

With reference to the disease germs in milk and domestic animals, and the great liabilities thereto owing to the economic character of our agricultural operations, little comment is necessary here, the facts being well known. Dr. Carpenter estimates that 80 per cent. of the meat in our metropolitan markets is unfit for human consumption. Despite all these and many other drawbacks, so great is our adherence to habit that we will butter our bread with margarine made from the vilest of fatty filths rather than relinquish a luxury which, in the ordinary ration, has no nutritive value whatever, and, to the unvitiated palate, is not even a flavouring substance. The area of land wasted in attempting to prevent adulterations —against which scientific rations are proof, as well as the area wasted in vain attempts to prevent contagious diseases amongst our domestic animals—is enormous, and has only the advantage of enlivening trade and affording respectable employment for a miserable pack of office-seekers.

A word is here necessary with reference to the pretensions

of our economic vegetarians on the ground of economy. They assert that they can live on "6d. a day" by the adoption of their rations, and parade this as an argument in order to win the patronage of the poor. If I were asked if I would recommend my scientific diet for poor people on the ground of economy, I would emphatically exclaim *No!* There is no inconsistency in this piece of advice. Under economic rule—the tyranny of monopoly—wages would irresistibly converge to the 6d. focus, and the poor would then be in a worse condition than they are now; for when the lowest limit is once reached there is no possibility of further economy; whereas, so long as the belief generally prevails that the subsistence standard is, say, 18d. a day, public resentment is aroused when a reduction is proposed, and at the same time opportunities are afforded for economising to bridge over temporary difficulties. A few amongst the poor may gain a temporary advantage under the 18d. standard, while no such advantage could be reaped under the 6d. regime. It is, therefore, the policy of the poor to maintain the 18d. basis, although they could otherwise gain a permanent advantage by living on 6d. a day. If the hundreds of thousands of unemployed poor in London could be educated to live on a porridge composed of a mixture of dust and rain, say street-scrapings, their condition would not be one whit ameliorated. So long as these scrapings could be obtained free of charge, they would not form a portion of our national wealth, so that a monopoly would have to be granted empowering some corporation to make the scrapings scarce by exacting, say, 3d. a day from all consumers of street-scrapings; then our national wealth would be enormously increased, and wages would sink to the 3d. level. The advantages in granting this monopoly would be that respectable employment would be furnished to police-officers and other officials for the purpose of protecting the scrapings against the assaults of hungry men, and the powers of the State would be invoked to guard the interests of the monopolists, thus strengthening the temporal and spiritual powers, and creating respectable employment for lawyers, judges, politicians, &c. It would also give employment to the unemployed, who, without monopoly, could have lived without labour, beyond the exercise required for gathering their food. It might also be the means of stimulating philanthropy. If some means could be devised for granting the monopoly to a corporation of poor people who had the power of making the scrapings so scarce that they were eagerly devoured as a luxury by the rich, this would be in a different category.

The central point in this chapter is the adjustment of the population to a given area, and so far as this principle is concerned, science may be dispensed with altogether ; or even should the science be proved to be false, the truth still remains that the extent of population is based upon the area of land which each inhabitant draws on for his support, no odds whether this area be one acre or one thousand acres. The space must be adjusted to the individual before the number of individuals can be determined. This is the law and gospel of population, which axiom cannot be reasoned away by all the abstract theories that have ever been conceived.

This discussion opens up two questions, namely, that our isles are not populated up to their full capacity—either (1) because the individual does not desire to till the area of land required for his support, or (2) there is some barrier which shuts him out from opportunities of cultivating his parcel of land. These questions will be solved in the two following chapters.

CHAPTER V.

Axiom III.—*In the opportunities for gaining access to the means of subsistence, no individual gains an advantage by being born before another.*

Deduction :—No generation can gain a similar advantage by living before another generation.

The weakness in the violation of this axiom is, that it leads to all manner of confusion.   What individual, corporation, or generation is entitled to this monopoly?   Who can present the most rightful claim?   Before what tribunal can his claim be legitimated?   That the title-deeds of our monopolists have been recorded in heaven need not be disputed; in fact, this is the weakest pillar in the whole structure.   The fabric of " divine rights " must fall by virtue of its own rottenness, and with the structure vanishes also the function.   In our own case, monopoly is still sustained by the theory that God was on the side of William the Conqueror and his generation, in the eleventh century, and their "heirs and assigns for ever." The effect of this theory has been the dethronement of the Norman's Deity and the crowning of a new Deity who presides over humanity, and defends with loyal impartiality the rights and titles of all generations of men.   An attempt, it is true, has been made to crown a Deity whose worshippers are restricted to the living generation in each age, *i.e.*, each of the living has an equal share in the monopoly—but the throne of this Deity is now a keg of fizzing dynamite.   That any Deity can enjoy peaceful and perpetual rule save He who presides over the interests of humanity is based upon the theory that any generation can bind and rob posterity.

Posterity may be stopped by the persistency of economic or other objective forces, but perpetually bound, and plundered, *never*. Our attempts to attain this end through physical violence, and through the moral compulsion of ancestor-worship, are the barriers which obstruct the solution of the social problem. The fact that the privileged individuals have received their titles by inheritance does not alter the situation, for there must have been an original monopoly; and to say that a title is just because it has been acquired by force or fraud is an acknowledgment that the reputed owner is a mere tenant, for he can only hold his title so long as he retains his superior force or cunning, or so long as he can bind posterity. Indeed, the so-called ownership is an abstract idea. The theory of "divine rights" is weak on the ground that the Deity, one would think, would enact laws for the preservation of his worshippers rather than for their extinction. The question really resolves itself into this: Can man by virtue of his supernatural powers of objectivity create a natural law through which monopoly can be perpetually enforced? Without denying the legal rights of the privileged successors of William the Norman, I shall show that there is no law in nature by which these rights can be maintained. The dualism is not that of peasant proprietary *versus* land-lordism—*i.e.*, little landlords *versus* big landlords—but between monopolism and natural law. It is the same dualism over and over again—man against nature. The solution of the land monopoly problem is the solution of the monopoly of all commodities; for everything, even man himself, is land or a part and parcel thereof. In short, land is the ground of all social problems, all social phenomena; and if the land monopolist does not own the human beings which his soil has produced, then his ownership is an abstract idea, and a scientific basis of ownership must be found.

For convenience' sake, all monopolies may be divided into three classes: (1) land monopoly, (2) monopoly of the products of the land, and (3) monopoly of men's minds and consciences. In the ultimate analysis, however, there can be no monopoly except that of the land; but if it can be disproved that our minds and consciences are products of nature, these being supernatural gifts, my classification falls to the ground. I am obliged to accept the above classification because my inquiry is scientific. If the land must of necessity belong to humanity, how can any individual or generation acquire a right to obtain or utilise its products except for the benefit of humanity? Here we are also confronted with a problem with reference to the rights of labour.

In order to remove all doubt with reference to the rights and titles of our monopolists, it is necessary to take a survey of the origin and development of monopoly. Hand-in-hand with this inquiry is found the monopoly of the human conscience; and the struggle for individualism in the one case runs parallel with that in the other. It is only a modified form of false subjectivity and objectivity—a shifting from one abstract basis to another. The form of individualism established by peasant proprietorship has no more to do with the solution of the social problem than the liberty granted to the individual for the control of his own conscience.

To arrive at the real origin of monopoly we must trace ourselves back to the "dumb brutes." Who has not heard of a dog monopolising a bone, or a monkey monopolising a stone to crack nuts with? In this inquiry, however, I can barely go back to the first page of human history. Indeed, so far as this chapter is concerned, man's original sin has its essential features in feudalism. This is the form of objectivity which has given rise to modern civilisation.

In very ancient times it was the custom of kings and other ambitious adventurers to engage in hostile expeditions against neighbouring States; and the fame and booty, the latter being largely in the form of land, were shared amongst the victorious leaders. In feudalism proper, which was built on the ruins of the Roman Empire, the system developed into theories of rights and titles, and these theories are the ground of modern laws in all civilised countries. This system having naturally brought about its own destruction, the ancient State, at a later period, did not recognise individual rights or personal freedom. The serfs, who enjoyed no share in the State, performed the menial and servile duties; and the whole life of the community—material, legal, moral, religious, &c.—was under State control. Ancient Rome, transformed into feudal States through the power of the Church and of the Teutonic race, had devoted its energies objectively to the arts of law, politics, and arms; and Christianity, the subjective element, had never succeeded in reducing the State entirely to an ecclesiastical institution—hence the development of a dualism which, for many centuries, divided the human mind and caused much suffering and bloodshed. The Church controlled man's spiritual nature, as the State controlled his outer life; the former by moral processes, the latter by outward compulsion. Rome remained the spiritual capital of the West after the political power of the Empire had departed; the universal State was transformed into the universal Church.

But the Teutonic character gave a different tone to the mediæval State. Unlike the Romans, the Teutons were not a political people. Individualism being their essential virtue, they could not tolerate the dominance of federated authority; they thirsted for personal freedom, and claimed to have innate rights which the State was bound to protect. The State existed for them, not they for the State; and the individual even claimed the sacred right of resistance against State sovereignty when his private rights were assailed or imperilled. A weak federal authority might therefore be justly anticipated, even though the Teutonic race cherished the Roman civilisation, and, above all, the Roman religion. In the mediæval State the supreme sovereignty, temporal and spiritual, was derived from God, and the temporal ruler being the head of an indirect theocracy, unity of creed was demanded, the clergy stood above the laity, thus enjoying many immunities and privileges, the peasantry still remained serfs, and custom was the chief source of law.

Under feudal monarchy, the land, as well as the political power, was conferred by the king (lord paramount) to his vassals (feudatories, lords, dukes, counts) in consideration of oaths of fealty and homage; and the lords were equally bound to guard the interests of their inferiors. The land allotments (feuds, fees, fiefs) were, upon non-performance of the stipulated military services (usually forty days in the year), and upon desertion of the lord in battle, to revert to the crown. But the system developed into a chain of tenants, those holding land from the crown being called tenants-in-chief, who again gave grants to middle lords, and these again to other middle lords, the lowest in the series being the villeins or small farmers who tilled the land with their own hands, were bound to the soil, and were in the most abject condition of servitude. These serfs dared not leave the estates without their lords' permission; they were in the same condition as beasts or other chattels, and could be chased off the land without a moment's warning. Those countries which adopted the feudal policy attained great military strength, so great that the people of other countries who enjoyed free tenures were obliged to surrender their allodial possessions, and then take them back under the feudal plan. The feud was originally held at the will of the lord, who was the sole judge as to the faithful performance of the vassals' duties and services, but afterwards it became certain for one or more years, and still later during the lifetime of the feudatory. It was not yet hereditary, although granted to the children of the vassal, until it became uncustomary to reject the heir, providing he was capable of performing the required services, so that infants, women, and monks, being incapable of bearing arms, were ineligible as feud-holders. The heir, however, according to custom, was obliged to pay a fine (relief) to the lord, thus re-establishing the inheritance, and even when the feud became hereditary the relief continued, although the original cause had vanished. It was a function of the feud that the vassal could not dispose of, exchange, mortgage, or devise it without the consent of his lord, for he held it by virtue of his personal abilities, neither could the lord transfer it without the consent of the vassal. The rent paid by the vassal to the lord was in the form of military services; that paid by the tenant to the vassal was usually in the form of corn and cattle, money having played an inferior *rôle* under feudal monarchy.

There arose also dualism between the feudal lords and their vassals. The latter, true to the Teutonic character, struggled for independence and supreme authority within their own domains, which resulted in reducing federal authority to a political fiction. The vassals, it was held, were not bound to serve the king, but only their immediate superiors or lords; and if the latter waged war against the king, their vassals could render them assistance without being guilty of treason. These rights were measurably the cause of political disintegration in the French and German monarchies. The rights of coinage and the administration of justice were also wrested from the king.

Since the tenth century, however, the wars were fought by the inferiors, the nobles, State-officers, &c., having gradually shaken off the military burdens imposed upon them by feudalism, which brought the system into great disrepute amongst the masses, whereby new feuds were rarely established, and elapsed ones did not revive, and war-duties gradually assumed their modern phases. Feudal jurisprudence, however, became a subject for the learned, and developed into that wonderful and intricate science (!) commonly known under the name, style, and title of "landlordism," which has given and still gives such high dignity to the academy and the bar. By the laws of Prussia (1850), the system became so obnoxious and demoralised, so shocking to the moral sense, that large tracts of land were expropriated without compensation to the landlords, and peasant proprietary reigned in its stead.

The feudal system existed in England before the Norman Conquest; but William the Conqueror recommended its continuance as the best method of strengthening the military position of the kingdom; and, accordingly, at the Council of of Sarum (1085), the leading landowners submitted to this system of military tenure. In the other aspect of feudalism, however, namely, the weakening of the federal authority, he instituted a marked contrast to the Continental system, the royal power being much stronger in England than in France. Most of the English estates were granted to William's followers as rewards for their military services. The English nobles have never enjoyed the rights of coinage or of private wars, and the mere possession of land confers no political authority, except in the abstract sense. The English theory is that everybody is liable to render military service, although our soldiers are drawn from the voluntary list. Many choose the soldier's life because of the scarcity of remunerative employment in other pursuits, caused by our economic system. If the feudal theory is sound, namely, that the fighters should own the land, our soldiers should, in the natural evolution of the system, be the landlords of to-day. Instead thereof, the people, by virtue of their being deprived of land, are forced into the army. During the Middle Ages the lords possessed both the wealth and physical endurance, while nowaday, owing to physical degeneracy, monopoly is maintained by the strength of political cunning.

Although the Wars of the Roses and the shocks of subsequent centuries may be regarded as having struck a fatal blow at feudalism, yet its shades haunt the economics of the present day, and uphold feudal law and politics with their ancestral associations and complications. Even dismissing natural and moral laws, the laws of England recognise no such title as absolute ownership in land. In the most absolute estate, that in fee-simple, the nominal owner is tenant in relation to somebody else, and so on up to the sovereign whose titles are not of this world. Neither is it possible for absolute ownership to take inception in any individual or any generation, for no tribunal can be constituted which is able to rob posterity of its birthright. Under our present law, when freehold tenants die heirless and intestate, the land escheats to the next lord, and in the absence of a middle lord, to the crown. When these are the results with reference to the highest estate which can be held in land, there is no use in discussing inferior titles, such as copyhold, in which feudal incidents are most marked, or other tenures which have arisen from inferior kinds of feudal holding. Herbert Spencer recognises land titles as being traceable to "violence, fraud, the prerogative of force, and the claims of superior cunning." A writer in the *Encyclopædia Britannica* (under Land), says : "They (the lawyers) expressed then as now the unquestionable legal rule that there is no such thing in our system as an absolute private right of property in land, but that the State alone is vested with that right, and concedes to the individual possessor only a strictly defined subordinate right, subject to conditions from time to time enacted by the community." Under Landlord and Tenant, the writer in the same *Encyclopædia* says: "No such thing as the absolute private ownership of land is recognised. The absolute and ultimate owner of all lands is the crown, and the highest interest that a subject can hold therein is a tenancy. The largest estate known to the law, that in fee-simple, is after all only a holding in which the owner of the fee stands to the lord in the relation of a tenant." A writer in a recent issue of an influential magazine[1] remarks: "The great landholders considerately in their own interests abolished the feudal tenures and created themselves owners," and then proceeds to say that "the monarch, representing the State, is still by law the feudal superior over the whole land of the kingdom." Those Irish landlords who trace their titles to the Cromwellian conquest, when confiscated tracts of land were awarded to Puritan adventurers, have no better titles than our feudalistic landgrabbers. All these tenures are as much of a fiction and an abstract theory as tenancy at sufferance in English law. Amongst the various offences through which "lands, tenements, and hereditaments" may be forfeited, those of crime and waste alone would be sufficient to dispossess the so-called owners. Guizot says: "Feudalism is perhaps the only tyranny which men have never been willing to endure: they have willingly yielded to monarchic and theocratic despotisms; but to feudalism, never." We have, however, willingly yielded to theories which are based upon feudal despotism. The cause of our not being willing to yield to feudalism or landlordism is very easily explained; for we can directly see that it shuts us out from the means of subsistence, and this explains the natural law which brings about the destruction of the system, the monarchic and theocratic despotisms having accomplished the same results in a more indirect manner; and their doom is none the less certain, although more remote. If our taxes were drawn directly out of our pockets, instead of indirectly, our revenue collectors would stand on the same footing as our feudal lords, and yet

---

[1] *Westminster Review* for May 1889, p. 468.

our taxes are much more burdensome by virtue of the fact that we have to pay them indirectly. So it is with our landlords; they frame schemes for relinquishing land which they do not and cannot own, and yet draw revenues from other sources, thus increasing those burdens which are already too grievous to be borne. To let go a monopoly in land and then seize a railway or any other monopoly, material or moral, is no advance towards the solution of the social problem.

Although the fact is recognised that no original titles have ever existed, yet it is said that innocent purchasers, having given substantial value, have acquired legal rights and titles of which they cannot be dispossessed without adequate compensation. There are three weaknesses in this theory. 1. If our feudal holders cannot retain possession of the land, the fact that consideration has been given can alter no natural law. 2. In the form of peasant proprietorship, the theory has been tested over and over again, and has just as often proved a failure. 3. If a sale of land can be enforced, the plain inference is that the ownership is an abstract theory.

Granted that the feudal holders had paid the value of the land a thousand times over, their possession would be no more secure than it is now. The natural law is that the means of subsistence ultimately fall into the hands of starving masses, no odds whether the titles be of this world or emanate from heaven. If the monopolists choose to bring about this state of affairs, they have only themselves to blame. Of all the forms of land tenure, that of peasant proprietorship is probably the most iniquitous. It creates the insane illusion that the occupiers of the land are the owners, and the " magic of property " incites them to the most aggravating form of slavery—a condition under which many a slave of the chattel class would succumb. The most abject and enthralling form of servitude is to be the slave of economic freedom. In France, Russia, and Prussia, even though the system is still in its infancy, no greater slaves have ever lived than the peasant proprietors, especially the peasants' wives and daughters. In Prussia, the peasants for the most part live in their cow-stables—perhaps it would be more correct to say that the cows dwell in the farmers' houses ; and in Russia, although the system is not much over a quarter of a century old, the land-plots are so minutely subdivided amongst the heirs, that their existence has become barely possible, and the surplus population, flocking to the towns and cities, has intensified the struggles for existence. In France the peasant proprietors are in a deplorable condition, although, owing to their increase of political power and the levelling down of the classes, their plight is not so appalling as it was before the breaking out of the Revolution. All the arguments which have been urged in favour of peasant proprietary, by illustrating the improvement of the French peasants, have a false basis, for any amelioration is due to the restriction of population rather than to the land system, coupled with increased economy in the administration of public affairs. The law of evolution under peasant proprietorship is, that the little plots of land, from which the barest means of subsistence is drawn, eventually find ownership after being mortgaged to their utmost value, in the person of money speculators, large farmers, retired merchants, &c., many of whom are the most exacting of all breeds of landlords; and landlordism, in the most iniquitous of all its phases, again holds sway. If it were possible for peasant-tenure to gain a firm footing, it would be followed by one of the greatest disasters which has ever befallen civilisation; for then the landlords, owing to their great numbers, could, by united action, forcibly deprive the rest of the community of their natural opportunities and their heritage, an occurrence which would not be feasible under the rule of a limited number of proprietors. The great political question in England to-day is, Who should be evicted—the landlord or the tenant? Science answers *both*. The mother-earth, which gave us birth and maintains our existence. belongs to humanity; and the individual—or even the generation—can only hold it for himself, his " heirs and assigns for ever," so long as he can stop or rob posterity, and exercise his miraculous powers of objectivity. The axiom that the land belongs to humanity is not only a question of common sense and justice, but also a question of natural law which we are obliged to recognise before the first step can be taken towards the solution of the social problem. We are coerced to acknowledge the supremacy of humanity. It is not a war of the classes against the masses, or the masses against the classes, both of which forms of warfare have failed to restore unity, law, and order: the problem of objectivity and subjectivity must be solved—we must know whether man, or some men, is a parcel of nature or an efflux from the skies.

With reference to compensation, there is no law in nature by which an individual, corporation, or generation can escape the responsibilities of his or its own follies and crimes. He who purchases land, either the raw material or the finished product, in the belief that he, for himself, his "heirs and assigns for ever," can enjoy quiet possession, must certainly run the risks attached to such a belief. Who compensated the Scholastics when the irreconcilable device of preaching Jahvism in the Church and Jovism in the Academy was resorted to? Who rewarded our Established Church when Methodism purloined vast flocks from her folds? Who compensated the coachman when the railway swept his tardy wheel from our highways? Who compensated the poor house-wife when cruel spinning-jenny drove her from her post? Who compensated Napoleon for his miscalculation on Waterloo's bloody field? Certainly he could not expect compensation from a *neutral* power, and if England had generously compensated him, what would he have done with the money? Matured fresh intrigues for the disintegration of the British Empire, of course. Who compensates the manufacturer of toothpicks for losses due to the rapid spread of vegetarianism in England? Who compensates the bears on our Stock Exchanges when the bulls gore up the price of stocks? Who compensates Christianity when her pillars are shattered by the battering-rams of Positivism, Agnosticism, Theism, and Humanitism? Who compensates the "boomers" of town property when in human desire the prices are regarded as being too fabulous for safe investment? From whose pockets are our politico-economic philosophers, our doctors, our lawyers, and our priests to draw their compensations when their occupations are gone? Who compensates posterity for the crimes perpetrated against it by the pride, folly, and

stupidity of past generations ? Who compensates our nation
when the shattered pillars of our museums and temples are
transported to distant lands as monuments of economic
barbarism ? The king who preaches that his regal power
emanates from heaven, but acts as if it emanated from hell,
loses his crown : the position is untenable, illogical—eternal
truth must prevail—and yet he blames the fulminators of discontent for his misfortune.

If our lords of the soil deserve to be adequately compensated,
they must be able to draw the same revenues as before, both
for themselves, their "heirs and assigns for ever ;" they will
require the same acreage of land for their support, and the
same toiling millions must wear their lives away in maintaining
ranks and ceremonies. No condition is changed; the privileged classes would draw interest instead of rent, which is a
mere change of words, and a change in the meaning of all the
words in our language would not solve the social problem.
If a compromise be once conceded as being just, this acknowledgment is an absolute relinquishment of all claims.

Let us, again, suppose that the landlords should compensate
the community for the loss of its rights and titles which it
has sustained for so many centuries; then the land, including
its products, must be either equally divided, the landlords
getting nothing, or some individuals must present their claims
for larger allotments or shares than others. How is it
possible to decide these claims, or even divide commodities
into shares proportionate to the weight or intelligence of the
various individuals ? What tribunal is to decide the extent
of their wants or necessities? If labour is to be the measure
of our rights, then who has the right to perform the most
labour? If war is to be the arbiter, then the history of
three thousand years proves this method to be a failure ; and
the condition remains the same whether the arbitrament is by
the bullet or the ballot. Who are the greatest criminals, the
robbers or those who have permitted the robbery to take
place before their eyes, when they enjoyed the power to
prevent it ?

In the previous chapter we found that population is a land
question—a science, not an economic theory. In the subse-

quent pages we have also traced the science of human rights to our mother-earth. Land is the root of economicism and of civilisation. Abstract theories regarding the land question have pulled us down; a scientific land basis is the only thing which can raise us up. The existing theories of rights, founded upon the laws of ancient Rome, have been based on the hypothesis that land tenures can by process of human law be made absolute—that man, by virtue of his supernatural gifts, can overrule the laws of nature. But the preposterous nature of this assumption can only be fully realised in connection with economic theories. A title can easily be made absolute so long as land is abundant, for when nobody desires land there will be no dispute about the title or character of the tenure. In order that land may be categorised as wealth, however, it must be made scarce; and when it becomes so scarce that everybody wants it, and is ready to sacrifice all his possessions for it, then the economic wealth of the nation, so far as the land is concerned, attains its maximum value. Thus a close connection between wealth and population is established; the greater the population the greater the aggregate desire; and the pressure of population against monopoly, titles, tenures, and vested rights, not against the means of subsistence, as the Malthusians contend, is the source of all poverty, misery, and crime. History proves that the monopolists, having feasted on the sweets of idleness and licentiousness through the toil and penury of the masses, drive this dualism to its appalling extremity; hence the havoc with rights and titles and the consequent disruption of civilisations. This is the objective extremity; then recommences the rule of the subjective element, the sentiment of humanity, the socialistic regime—the bond with which nature endows all animals for the preservation of their species; the subjective element then again evolves into the objective, and so on in ever-recurring cycles.

But this is not yet the logical conclusion of monopoly. The economic value of land is not restricted to the desire for it in any given generation. The prospective increase of population gives value not only to unoccupied lands, but also to land in the centres of agricultural and commercial activity. Thus we find that in our economics we are ruled by posterity, while in our morals we are ruled by the shades of our ancestors, the living generation being, as it were, the battle-ground for the opposing factions. We are amid the smoke and clash of arms between the dead and the prospective living. There is a saying, uttered with the view of tickling our laugh-faculty, to the effect that we owe nothing to posterity because posterity has done nothing for us; but this is a question of the greatest gravity in the solution of the social problem. We coerce posterity to create our wealth, and then rob it not only of the means of its subsistence, but also of the strength which is required to obtain its livelihood; and then we invoke the shades of our ancestors, under the sacred name of religion, to sanction our crime. So violent a dualism undermines our whole fabric of economics and jurisprudence, and transforms our entire political system into a myth. My position, therefore, is to find a scientific basis of land tenure, which must also include the science of human rights.

With this chapter we started out with two propositions: (1) the barrier which hinders the individual from obtaining possession of the parcel of land required for his subsistence, and (2) his desire to cultivate his plot. This barrier, as we have just seen, is monopoly; and it is now in order to discuss the second proposition, which will be further elaborated in the next chapter. The collateral issues, namely, the nature of the tenure, the alternative of an equivalent return should the individual not desire to cultivate his plot, and the extent of his rights to labour on other plots of land for the purpose of treasuring up wealth for himself and his "heirs and assigns for ever" will be discussed in subsequent chapters.

We have seen that in ancient times the tillers of the soil were slaves, in mediæval times serfs, and that the lords of the earth have always been the landowners, rulers, idlers, and slave-drivers. In all ages brute force was considered to be the essential in growing and gathering the fruits of the earth, and the occupation therefore only befitted the most benighted slaves. With the rapid increase of population it became impossible for the lords to augment their numbers in propor- tion to the general growth of the community, and serfdom

being a revolting condition, an influential middle class emerged
into history, who naturally applauded any theories which
tended to procure for them a *respectable* livelihood, if not
also rank and affluence. Any theory which aimed at the production of wealth with the view of maintaining an increased
population flattered the lords, because the pressure of the
people against monopolised land was found to increase rents;
and so long as the lords were regarded as absolute owners of
the land, the tenants being mere chattels, population was
encouraged, the masses being the consumers of agricultural
products and the producers of luxuries for the lords, which
developed into the autocracy of *fashionism*—a magic power
which has permeated all classes of the community, and is a
convincing proof of our Simian origin. Without the rule of
this Czar, who is a deadly foe of science, economicism would
be an idle tale, and wealth would be extinguished. His
rights are rapidly growing divine, and he requires many
million acres of land for his support.

These two barriers in the way of land cultivation, namely,
(1) monopoly, and (2) the idea that agriculture is a servile
occupation, are all that is necessary to be discussed here.
But when all the land in a country is monopolised and occupied there is no way of measuring the strength of the desire
to cultivate any portion of land. In all sparsely settled
countries, however—America, for example—where there are
millions of acres of free homesteads, we find that the desire
to occupy and cultivate land is very weak, even amongst
those who are able to undertake agricultural operations; and
still weaker is the desire to embark as agricultural labourer,
though the wages be higher and the means of subsistence
securer than in other manual occupations, despite the fact
that the skill required may be less. The dread of the stigma
which is attached to agricultural pursuits, and of the social
degradation conjoined therewith, are the chief impediments.
Stock-raising, however, which, as we shall see, has evolved
into the dignity of a commercial enterprise, has opened up
employment for people who are satisfied with a moderate
degree of respectability. The respect due to trade has become
so venerable that the peddling of trinkets is preferable to

agricultural serfdom. The holding back of lands owned by huge monopolists from cultivation, thus driving the pioneers beyond the borders of the community, out of the reach of social advantages and enjoyments, adds greatly to the iniquity. So firmly is the sacredness of land monopoly rooted in the civilised mind that the monopolists, although their wealth is due to the increase of population and the sums squandered by the State for the importation of labourers, demand compensation when public indignation is aroused, and when the legality of their titles, acquired by all the fraud and cunning incident to the lobby, is called in question. The titles wrested from the legislators in this manner have no greater validity against humanity than those acquired in the most crime-polluted wars which have ever disgraced the pages of history. Ignorance of natural law is not less excusable than ignorance of human law; and the landlord who believes that lobbyistic is more sacred than the feudalistic title must run the risk of such belief. If he imports immigrants to cut his throat, no odds whether they belong to the free or the chattel order of slaves, he only repeats what our mercantile theorists did when they gave encouragement to population. The landlord who permits his hogs, by virtue of their numbers, to outwit him in strength, must run the risk of their breaking into his corn-fields, especially if he attempts to subdue them by starvation. In the United States the negroes are already threatening the whites with extinction, due to their superior hardiness and fecundity.

The evolution of slavery is so distinctly marked that it can give rise to no misconception. First we have the chattel slave, second the serf slave, and third the economic slave, and yet we call it the evolution of freedom. The chattel slave can never be in a worse condition than other domestic animals; for it is the personal interest of the owner that they should be permitted to draw on sufficient land for their support, otherwise their efficiency could not be maintained. Not so, however, with the serf or the economic slave, for his efficiency bears no relation to the profits of the slave-driver; and in this respect the economic or free slave is in a worse condition than the serf. The true measure of the intensity of slavery is

the area of land on which the slave is permitted to draw for
his support; the more this area is below the standard for subsistence, the more intense is the condition of slavery. What
is the sense in talking about human rights or political freedom
unless these so-called privileges enable the toilers to draw on
a sufficient area of land for their subsistence? We boast of
the spread of education amongst the masses, and regard this
as an indication of progress, prosperity, and freedom. What
are they educated to know? They are taught to know that
wealth is produced by creating lords of the soil, thereby
making land scarce; that in this manner capital accumulates,
which is the fund from which labourers must live, and is also
the source of employment; that they must support spend-
thrifts to waste the products of their labour, thus creating a
greater demand for labour and encouraging trade; and that
when capital tends to decrease they must accept lower wages
and restrict their numbers owing to over-population. They
are taught, moreover, the sacredness of "vested rights" in
ideas. The learned depend for their subsistence largely upon
accumulating the ideas of the past, and their livelihood depends upon this education just as much as the landlord's
depends upon monopoly. No attempt is made to show any
relation between these ideas and the welfare of humanity;
and the masses are taught to believe that men of learning
also must exist for the purpose of consuming the products of
their labour. Any attempts made to render the study of
these ideas unfashionable are resisted by men of learning, just
as landlords resist attempts to disturb monopoly, and in the
abstract method of thinking, wealth is destroyed by extinguishing a profession as well as by extinguishing rent.
Hence the theory of vested rights is applicable to our intellectual as well as our material pursuits; and touching respectability, if we cannot reach its height by supporting ourselves
without labour, we must rush to the next degree by maintaining ourselves through mental exertion. The masses are
taught to believe that if any one or more of our intellectual
occupations were removed, there would be greater competition in the labour markets. Here the source of the degradation of labour and the prominence given to intellectual over

physical employments may be traced. Our religion binds us to believe that the spirit is the pure, undefiled, immortal part, while the body is a mass of corruption, and our philosophy teaches us that the idea is the absolute and the sublime. The high degree of respectability attached to the idea is based upon the theory that the function remains when the structure is gone, and the prominence given to the idea leads to the conclusion that ideal theories must be relied on for the solution of all phenomena, natural and supernatural. The man who bases his titles to the means of subsistence upon ideas must run the same risks as other clamourers for the sanctity of vested rights, his endeavours to bind posterity being equally futile, and unless he gives adequate return for the area of land which he draws on for his subsistence, no power or law on earth can save him.

There being no law by which an individual can hold land by ordinary metes and bounds, another basis of tenure must be found. One thing is certain, namely, that every individual draws on a certain area for his support, no odds who claims to be the owner of this parcel of land; and it makes no difference whether the area be located in one spot with definite boundary-lines, or split up into minute parcels and located in every country under the sun. This is not a question of any scientific import; for the area may be conceived as being definitely located, and its products exchanged for those in other districts or countries. All lands are measured or valued by their products. To deny that an individual has a right to draw on land for his support is to deny his right to live; and if he has a right to live, he is the real owner of the land from which he draws his subsistence. There can be no meaning in claiming a title to a piece of land so long as another person has the power to appropriate the products. The economic idea only leads to abstract conceptions of rank and respectability. While it is true that the abstract title gives the holder power to draw on land without rendering adequate service to humanity, yet the holders of no titles may draw on still larger areas without claiming any land titles whatever, and without rendering corresponding service. Science has only to do with the area of land which an individual draws on for his subsistence. All the chattel slaves who ever lived, in common with all other domestic animals, have enjoyed the area of land required for their support both for themselves and their "heirs and assigns for ever;" and as this tenure has never been disputed, it is as absolute in human as in natural law; and the economic slave, as heir to the property of the chattel slave, still enjoys the same tenure, unless debarred by some statute of limitations. By virtue of this law, nobody can legally suffer for want of sufficient food, clothing, or shelter, and all our charitable institutions are therefore out of place. When man refuses to recognise justice, no pouring out of charity can avail. The recognition of line-fences or international bounds is as impracticable as it is useless; for every new-comer into the world, every new claimant for his share of the means of subsistence, would have the effect of changing every boundary-line and fence on the face of the earth, if Humanitism became an universal system; and every death would lead to the same complications, as would also the varying quantities of food consumed between birth and decrepitude. So long as each individual draws the income from his parcel of land, it does not concern him where it is located. The right to draw on land, the means of subsistence, is inseparable from the right to exist—that is, there can be no line of demarcation between the area of land required to produce our bodies and souls and that required to maintain them in health and strength. If this proposition is not true, then what tribunal is to decide at what period in our life the millstone is to be tied around our neck and cast into the sea? Who amongst us is to be the first to undergo this ordeal?

This chapter would not be complete without a word about the land and conscience monopoly of the Church. Feudalism and the Church developed side by side during the Middle Ages, and the decline of both institutions is also conspicuous for their simultaneous transition into modern history. The one was a necessary counterpart of the other, both from the political and the moral point of view. Those gigantic land monopolists, the priests, often acquired their titles through other schemes than directly through the Deity. Much of their land had been originally granted to them by princes;

and they also derived large revenues from votive offerings or voluntary gifts, and customary incomes of various other kinds. During the expeditions of the Crusaders, they accumulated vast treasures; and fortunes were saved or acquired by saying masses in the families of the nobility. From the fourth to the tenth centuries the clergy were regarded as a privileged class, free from feudal burdens; but after the tenth century they became liable, except through purchased immunity, to military service for their estates. Their possessions thus being feudalised, the clergy were classed as vassals of the kings; but, as the Pope maintained, the Church was independent of the temporal power, which assumption gave rise to a series of struggles between the regal and papal sovereignties. The situation then became tripartite; for there also existed a struggle between the king and his vassals, as already mentioned. During the thirteenth century science and philosophy began to reassert their sway, which also fell heavily upon the Church. The modern transition is marked by Protestantism, science, revolution, free-thought, economicism, and the evolution of serfdom into free slavery.

The land titles of the Church, one would think, coming more or less directly from the Deity, were as absolute as can be imagined, and yet, by some mysterious working of natural law, the clergy cannot hold land more firmly in their grasp than our temporal lords. Tithes have been resisted with the same vehemence as other rents, despite all the terrors of boycotting—formerly called excommunication—and stringent laws have been passed for the punishment of those who so pervertently attempted to defraud the Church, as agent of the Deity, of her spiritual dues. The Church has lost all legal and spiritual title to her tithes; for there remains not a vestige of the ancestral tithe-principle, and Jahveh has long since refused to rebuke the devourer "who eateth and spoileth the fruit of your ground, and your vineyard shall no more be barren in the field," and to pour out those other abundant blessings due to the offering-givers and tithe-payers. One of the profoundest of all politico-theological problems is to determine how the royal crown and the sacerdotal stole, both emanating from the same Deity, became hereditary with respect to the former and non-hereditary with respect to the latter.

The tithe system, however, has had the advantage of furnishing respectable employment for lawyers, as well as ecclesiastical judges, free from the contamination of servile labour, and our politicians have received their compensations in the respectable and congenial employment of passing various Acts for the relief of the clergy and the punishment of those who were so unscrupulous as to dodge the payment of tithes. Despite the violent struggles which have taken place between Church and State, the clergy and the lords, they have formed a friendly bond to maintain their ranks and titles, to achieve monopoly, and to rob posterity. Land and conscience monopolies stand or fall together. Feudalism, so revolting to the moral sense, would have melted away like a snowflake beneath the summer's noonday sun had we not been bound in our conscience to obey our masters—" the powers that be," which are " ordained of God." When we are permitted to look behind the scenes—when we see that religion is no longer boxed up within coffin-walls and clothed in dead languages, that theology and law are no longer buttoned up in silken gowns, and that justice is no longer hatched on woolsacks—then will dawn the age of science and Humanitism.

Far be it from me to condemn the Church as having been a useless institution. The excessive objectivity which ladened Greece and Rome with decay and ruin, and the unnatural subjectivity which left the Jew without a clod whereon to lay his head, must be carefully weighed in the social balance. Feudalism itself, as well as Christianity, was a necessity of the age; the excessive objectivity required the restraint of some subjective element. The Church, although herself engaged in objective enterprises as daring as those of military adventurers, exercised a morally beneficial influence over her tame flocks, and inspired them to many virtuous deeds. She deserves our compassion for the good she has done in awakening our sympathetic instincts, but she equally merits our censure for her persistent efforts to perpetuate the structure when the function is gone. She preserved a spiritual bond amongst the nations after the political bond had been snapped asunder—a link which may yet be welded into a chain of steel for the embrace of all humanity.

The Catholico-feudal page of the history of our race will not go down to posterity as a landmark of human progress. The system is the legitimate parent of economicism, which will mark a retrogressive period—the transition from serfdom to economic slavery, and the equally pernicious imbecility of lordly pomp and demoralisation, with its tendency to harden our hearts and make us callous to untold agony and woe. It has imbibed no humane ideas with regard to human rights, but has intensified human wrongs. The transition to spiritual individualism, marked by the Protestant era, is merely proto-typed by the transition from big landlordism to little land-lordism—from vast estates to peasant proprietorship; and so far as science and humanity are concerned, it makes little difference whether the individual is his own priest and prophet or hires highly-toned and lettered functionaries to point out the way to everlasting salvation. Our rulers, the landed lords, have manipulated affairs for their own selfish ends, and have only succumbed from time to time under the severest pressure. It would be folly to expect a different policy even now. Attached to the land are hereditary castes and cere-monies without which the estates would lose their charm and great value. Let these customs disappear and the land titles also vanish; for feudal estates in land can only be made "absolute" by a return to the antiquated system in all its purity and entirety.

The assertion that the Catholico-feudal economic period of human history will mark a retrogressive era is, however, to be taken in the limited sense. With respect to universal evolution in the life of which this era is but a potent throb, a progressive impress will be stamped; for the knowledge acquired by the experience of the past two thousand years will enable us to escape manifold dangers in all the ages to come.

CHAPTER VI.

ECONOMIC AGRICULTURE.

HAVING pointed out the importance of land in relation to the social problem, no apology will be required for devoting a chapter to agriculture. It may be asked, however, what this subject has to do with abstract theories, or what right I have to classify it under abstract methods. If I can show that the mainspring of all our actions can be traced to the theories of ideas, no discussion can arise. It is sometimes said that science rules the world, which assertion is contrary to fact. We are controlled by economicism, and scientists are still, for the most part, economic men. They search that they may know rather than that they may act; or, so far as action is applicable to their researches, they are labouring in the service of economic men—even more so than the Scholastics formerly laboured in the service of the Church. Until science shakes itself from the fetters of economicism, just as peremptorily and effectually as it has done from the fetters of the Church, no step can be taken towards the solution of the social problem. There is no sense in knowing how it may be solved unless we turn our knowledge to practical account.

Agriculture is important to us in many ways. It is the occupation through which we obtain the necessities of life. Had man remained in his natural state, no other occupation would have been necessary for his existence, although a variety of other occupations could have been made conducive to the development of his structures and functions. The fact that agriculture is the basis of our existence should be sufficient to convince our economic theorists that an economic basis must have a short career. Pastoral employments remind us of the rural simplicity, with its attendant pleasures, which characterised the " wisdom of our fathers," and of the dignity and grandeur upon which poets in all ages have loved to dwell. They remind us of the fact that the philosophers of ancient Greece, so lauded to the skies by the abstract thinkers of modern times, imperatively condemned economicism, choosing agriculture as the basis of national prosperity and of human progress. These very philosophers, whom we delight to call heathens, saw " with that inner eye which no calamity could darken," the terrible consequences with which economic rule would be fraught; and although their predictions have been fully verified, yet we still continue to boast of our civilisation, and of the theories through which it has been achieved. These results should stimulate our theorists to dip more deeply into the fount of philosophic inspiration unsealed by their heathen brethren in ancient Greece.

It is out of place in this chapter to review the origin and development of universal agriculture. What we require to know here is the effects of economicism upon the modern agriculture, and for this purpose it is necessary to trace the evolution from pioneer times to the present day. This, again, involves another question, namely, the relation between primitive simplicity and happiness. Has human happiness kept pace with economicism? We speak about the increase of life's comforts, which is largely a question of temperature; indeed, the whole question of happiness or pleasure may be strikingly illustrated by variations of temperature, and the scientific aspect of the inquiry is, Should we accommodate ourselves to natural temperatures, or can we bend them into conformity with our artificial conditions? This is again the old, old story of objectivity and subjectivity. For example, in England— probably it would be more correct to say London—the most agreeable temperature is about 65° Fahr.; at 70° it begins to swelter, and at 60° we begin to shudder with cold, if we are not warmly clothed and muffled. On the other hand, the pioneer in a cold climate feels comfortable whether the temperature be at zero or 90° Fahr., and work does not become a burden to him at either of these extremes. The temperature of his house may range at least between 35° and 80° without his experiencing discomfort while he is enjoying repose, even without a change of clothing, and any sudden changes in the weather produce no deleterious consequences. Is it now not absurd to say that the Londoner under a temperature of 65° is happier or more comfortable than the pioneer under 55° or 75°? But we are not dependent upon the modern pioneer for an illustration of this sort. I may again be permitted to mention the name of Socrates—although I have been baffled in my attempts to ascertain whether he was a religionist or a philosopher—whom the Delphic oracle in ancient times, and the greatest of economic and Utilitarian philosophers in modern times (John Stuart Mill), pronounced to be the wisest of the Greeks—or, which is the same thing, the wisest of men. Let us now tear a leaf from the life of this wisest man of all the Greeks.

Socrates in the campaign at Potidæa exhibited a power of endurance unequalled by all the other soldiers in the camp. When provisions were short he fasted with the least complaint, and when a feast was the order of the day he ate and drank in the heartiest manner. During one night in a severe Macedonian winter there was a keen frost, and while the other soldiers stayed within or went out comfortably clad in sheepskin jackets and felt shoes, Socrates alone walked about in the open air, clad only in a common mantle, and, with his bare feet, trod the frozen ground more alertly than those whose feet were warmly sheltered. On another occasion he went out early one morning to engage in contemplation ; but, the object appearing very abstruse, he remained standing and looked straight before him till near midday, when the soldiers noticed him. Towards evening, some of the Ionians, bedding themselves in quilts and carpets, slept out of the camp with a view of keeping an eye on the philosopher, and when they awoke in the morning with the rising of the sun there he was found standing in the same spot; and having offered a prayer to the sun, retired.

With reference to the military bravery exhibited by Socrates, which was as great as his moral courage, nothing need be said here. All I wish to illustrate is, that happiness is a matter of habit, and can be attained without leading a life of economic indulgence. Economic happiness has a very limited range; it is the forerunner of pain in the indulger himself, and his condition is painful to all beholders who are moved by loftier aspirations. It is needless to repeat that the more limited the sphere in which a person moves the more land and labour he requires for his support.

In going back to pioneer days, I may confine my observations within the memory of men still living, who, in their early days, did not live under economic rule. I will select a territory in the colder regions of North America, which will bring the subject home to ourselves, because it includes a large portion of our greatest colony, Canada. I do not intend to insinuate that pioneer life is not still progressing in the more northerly and westerly regions, but they are now under the economic dispensation. It would be unnatural to suppose that our agricultural ideas are sound when we are so exceedingly unscientific in matters pertaining to economics, dietetics, and rights.

Although it is true that land monopoly preceded the pioneer on American soil, the most gigantic of all the monopolists being the State, and in Canada there were also the Hudson's Bay Company and the Canada Company, yet, owing to the vast areas of fertile land, and consequently to its low value, the effects of land monopoly could not be appreciably felt, there being many million acres of free homesteads as well as other lands which could be purchased for nominal sums of money. Land values then developed on account of prospective pioneers rather than of those who had already settled. These pioneers had to work hard, as everybody should do, but there was no poverty, and everybody enjoyed the fruits of his labour. As all the neighbours were on the same social level, there was nothing to mar their harmony. No laws were required, and each was always ready and willing to help all, and all each. The only professional gentleman required was the clergyman; and as for the merchant, he was far off, his functions being extremely limited. He was usually himself also a pioneer engaged in agriculture, and as economic ideas had hardly been imported, his sympathies were with those with whom he daily intercoursed. Indeed, had his customers from the "Old Country" not been

economic men originally, his stock-in-trade would have been
a miserable spectacle. These sturdy bushmen, and especially
their families, evaded colds, not on account of their scientific
knowledge of ventilation, but because the wind and snow blew
through every chink in the log-walls. Their home-spun garments were scanty, not from necessity but from choice, for the
wearers were capable of motion during their working hours,
and by their strong vital system they were able to digest
their food, thus chasing the cold away by the generating of
heat within their bodies. The children needed no boots even
in winter, so long as they were permitted to climb the snow-
banks, chase one another, or run races with the dogs after
the flocks and herds; and their hands needed no protection,
even while handling brush under the snow in the woods. In
summer, the boys and girls went to school with their lower
extremities, below the knees, bare; and yet they felt not
ashamed, for their virtue was as strong as their physical
powers of endurance. They needed no shoes while racing·
over gravel or corduroy roads, because sandals grew on the
soles of their feet. Should their bare hands become cold
during winter "snaps," they warmed them by casting snow-
balls at one another, or at objects along the way. In order
to enjoy their higher education, they often travelled three or
four miles to school, and even farther, the parents being the
teachers of the elementary branches. These pioneers needed
no lawyers, because there was nobody to foment quarrels,
and disputes about line-fences and title-deeds had not yet
begun; and as for the doctor, it is even disputed at the
present day whether he is an improvement on the pioneer's
granny. Indeed, these very grandmothers, if they made a
profession of their craft and drew professional fees, would,
owing to the limited number of their patients, have become
the first paupers known within the sphere of their activity.
The bushmen needed no flouring-mills, for they had teeth to
do their grinding, although the pepper-mill was also utilised
for this kind of work. They needed no sugar, because they
were able to masticate their food until its starch became converted into this dainty, and, failing this, they had abundance
of fruit full of juicy sugar. The sugar made from the sap of

the maple-tree was mostly for sale, although some families consumed considerable quantities of maple syrup. They did not need meat or butter, because the children were able to climb large forest trees for nuts, and also delighted in nut-gathering expeditions, although the carcass of a sheep or a hog, suspended against the house walls, often adorned the surroundings. There was nothing to mar the harmony of the neighbourly gatherings in the family homes, which developed into school-house clubs, because there was only one grade of society, and the era of economicism had not yet dawned. At the to-and-fro excursions during these social and literary gatherings, there was nothing to disquiet the stillness of the moonlit scenery except the jingling of the horse-bells and the mirthful voices of the youthful excursionists; and an occasional upset, spilling the occupants of the sleigh into the snow, was the culmination of all the merriments. In short, our pioneer heroes and heroines scarcely knew any object in life except happiness, and they possessed the requisite health and strength to enjoy it to their heart's content.

But these were the days anterior to the dynasty of economicism, and civilisation had not yet begun to dawn. Beyond municipal matters, little interest was taken in politics. Then came the politician clamouring for "vote and influence," and gave, by way of compensation, trinkets and kisses to the babies. The mothers whose babies were not so sweet or fortunate as to become the recipients of political gifts had no alternative than to buy dolls and other child-appeasing toys at the village store. In this manner the mother performed two righteous deeds: (1) she raised herself to the same social level as her bribed neighbour, and (2) she encouraged manufacture and trade. The village storekeeper, through expansion in the volume of his trade, could now afford to attire his children in doll-like fashion; and when they thus met the rustics in the village or country school, the latter felt so alarmed that they could no longer go to school without a fig-leaf to hide their shame. The storekeeper, maintaining the style of comfort observed in the city where his business duties called him from time to time, now becomes the centre of the social circle in his locality. Indeed, in the lapse of time, the doctor and the lawyer no longer considered it beneath their dignity to settle in the village; and although the inhabitants are still too sparse for their support in the style befitting their professions, yet their rural patients and clients are rapidly on the increase, so that the respectability of these professions is maintained. The village council soon begin to devise means for the increase of population; for, as absolute owners of town lots, as well as for professional and commercial considerations, they become convinced that this policy is a masterpiece of sagacity. But in order to attract population there must be something for the people to do, and high wages and high profits on investments must be promised. There not being sufficient clerkships to induce a material increase of numbers, there must be manufacture—which occupation is only a step or two lower in the scale of respectability than that of the merchant—and here is also an excellent opportunity for gaining the ear of the farmers. As there is no use of manufacture without trade, a linking with the world's commerce if possible, the council offers a bonus to any person or company who will erect a doll-factory, providing the farmers grant a bonus to a railway corporation which will carry the surplus dolls to a point which will connect the village with the world's markets. By this scheme three advantages are to accrue to the farmers. 1. The manufacturer of dolls will consume their butter and eggs at their own doors, thus opening up a home market. 2. The railway will carry said butter and eggs to the world's markets. 3. The development of the village into a city, caused by the manufacture of dolls, will offer opportunities for respectable employment for the farmers' sons; whereas, without such opportunities, the boys and girls will have to go west and take up land far away from the dear old spot which gave them birth, and all the lands within the borders of civilised communities being monopolised and held vacant for higher prices, the only resort is that the youths must settle amongst the wild beasts. The scheme sustained, the population of the village, now grown into a town, is increased by the importation of money-lenders and their agents, and the volume of law business becomes enormously increased. The rapid growth and prospects of the town, and especially the granting of additional bonuses, attract more merchants and manufacturers, more town-lot speculators, and a motley crowd; the farm mortgagee becomes an important and influential citizen, and the town, encouraged by more extensive business facilities, leaps into a city. All the inhabitants are taxed, directly or indirectly, to maintain the upward tendency of land values for the benefit of a few owners; journalists are engaged to "boom up" the enormous advantages which the city offers to enterprising and industrious men, and the vast agricultural and mineral resources of the surrounding country. Seventy-five per cent. of the farm lands is mortgaged to one-half of their value, partly to maintain this splendour, partly to educate the boys and girls for respectable occupations, and partly to maintain the style of living demanded by the circumstance of proximity to a city of such proud pretensions. Meanwhile, however, other towns spring up, which offer still larger bonuses, thus inducing manufacturers to leave the great city; the collapse comes, followed by the pernicious consequences due to all such catastrophes. Without a moment's warning, the wealth sinks from millions to tens of thousands—simply due to a freak in human desire.

But these consequences are merely preliminary to the more startling effects of economic evangelism. The farmer has evolved into an economic man, indulging in the usual luxuries, and the greatest ornament on his farm is the mortgage. Some of the boys have learned respectable trades or professions, and the girls are also rising in the social scale, for they have learned to make fashionable dresses for ladies and dolls, or to deal out articles by weight or measure over the counter. One of the boys, however, has disgraced the family by becoming a pedagogue; but this is only a temporary loss of respectability, for as soon as the farm can stand additional mortgaging, he will commence to study one of the respectable professions. The boy in the family whose ambition for respectability is least developed remains on the farm. Our locality has now developed into the second generation, the old pioneers having retired, leaving the farm management in the hands of the boys, and many of them have gone to that abode of happiness of which the clergyman has so often spoken, and about which he has so often preached. The respectable members of the family pay occasional visits to the old homestead; but as their aged parents, or their brothers and families who now conduct the farm, are much lower in the social scale, being also deficient in education, the visits cease to become enjoyable, and a rupture takes place in the relationship. But, social ambition now being aroused, better days are in store for the home-keeping youths whose wits have remained so homely. Having failed to pay off the mortgages by deforesting and cropping nearly all the land, thus depriving the crops of protection, causing the fountains and springs to dry up, and making the rainfall uncertain and irregular, as well as encouraging all sorts of insects injurious to the weakened agricultural plants, other methods of agriculture have to be resorted to. The soil becomes exhausted through many years of excessive cropping, and a school of scientific agriculture has to be established. The scientific professors having failed to restore the robbery of the soil, and of the vitality of the agricultural plants, by means of artificial fertilisers, grass now being the most natural plant, agriculture proper develops into stock-raising. The theory is, that stock manure is the only fertiliser which will restore the fertility of the soil; and the ancient motto, "More stock, more manure; more manure, more grass," is put into practice. The live-stock era is now inaugurated, and the "battle of the breeds" rages with great fury. The owners of "scrub" stock fall into disrepute, and the only respectable branch of agriculture is the keeping of stock with long pedigrees and fancy prices. The rise and collapse of stock "booms," which have developed into speculative methods, based on the soundest principles of political economy, have probably been more disastrous to our agricultural interests than all other influences combined. The tinkering with herd-books alone, the constant entries and erasures of so-called "spurious pedigrees," and the gigantic frauds connected therewith, accruing to the advantage of a few soulless individuals and corporations, are more than any legitimate industry can stand, and the false records of performances in the battle of the milking breeds carry their own doom. Some of these breeds are excellent at the start of their career; but through incessant stuffing and drugging they lose their vitality, engender disease, and then their race is run.

We have here struck another dualism, the solution of which is of vital importance in connection with the social problem. According to the agricultural theory, stock must be raised in order to produce manure; otherwise the fertility of the soil cannot be maintained. I am not here going to deny that chemical fertilisers have not always produced profitable results; but, as we shall see in a later chapter, our agricultural scientists have made exactly the same mistake as our orthodox doctors when they accepted chemistry as the basis of agriculture instead of biology—that is, they accept physical and neglect vital laws, and these apply to plant as well as animal life. Now, if we must raise domestic animals in order to produce manure, and, as we have seen, their flesh is not fitted for human consumption, then the question becomes dualistic, namely: 1. Must animals be raised merely as manure machines and their carcasses fed to hogs or utilised as fertilisers? 2. Must the putting on of flesh be the primary object, the manure being a secondary consideration, and the carcass consumed by human beings merely because it is cheap, not because it is fitted for food. In short, which is the basis—the manure or the flesh? If animals must be raised for manure, of course their flesh must then be cheap. This question, however, is the economic, not the scientific. Taking social conditions as they now stand, there are agricultural professors who maintain that the manure is the main consideration in stock-raising, and yet these very same authorities are in favour of the thoroughbreds. The pure breeds of stock are not defended because they are greater consumers than the so-called "scrubs," but because they yield better results from the same quantity of food. This is a ridiculous position; for if the results are better so far as beef or milk is concerned, they must be correspondingly less concerning the production of manure, for the same nutrients in the food cannot produce both beef (or milk) and manure. When manure is the basis, the professors are therefore obliged to accept the breeds

which are the lowest beef or milk producers, and when beef
or milk is the basis, they are obliged to accept the breeds
which are the lowest manure-producers. From the practical
standpoint, however, farmers always insist upon large yields
of beef and milk, and are so indifferent about the manure
that many of them do not exert themselves to save it. This
same question, moreover, is closely allied with the theories
of population. If, for example, a breed of domestic animals
could be developed which deposited as manure all the food
consumed, then, the manure all being saved from waste, the
soil might become so fertile that there could be little or no
limit to the population a given area could sustain, especially
if the manure-producing animals became so numerous that no
land could be spared for the raising of other animals. Possibly it would not be unwise for the Malthusians to take up this
argument in proof of their theories. Such utter nonsense is
quite characteristic of abstract methods of thinking, and is
scarcely paralleled by the idea that scarcity is the basis of
wealth. The absurdity to which the manure theory leads
could only have originated in the abstract conception that
man's life is sustained differently from that of other animals,
he being of supernatural origin.

This theory became so prevalent amongst agricultural
authorities and practical farmers that I deemed it my duty in
other writings[1] to expose its fallacy; and as my conclusions
have been accepted by some leading agricultural authorities
and disputed by none, it is not necessary to enter into detail
here. Besides, the discussion is too lengthy and technical to
be introduced here, so that I shall confine my observations to
the points which I discussed in the writings referred to as
follows:—1. Animals can return nothing to the soil which
they did not previously take out of it in their food, and the
returns in the form of manure must therefore be less than the
fertility removed by the quantity of beef and dairy produce
sold off the farm. 2. The quantity of produce sold off the
farm or not returned to the soil is the measure of soil-
exhaustion. 3. When the land produces the heaviest pos-

---

[1] Stock-Raising and Grain-Growing in relation to Soil Fertility and
Exhaustion.

sible yield of stock foods, the soil-exhaustion is much greater than can possibly take place through grain - growing; but when grain (*e.g.*, wheat) reaches the maximum yield, while the stock foods (hay, roots, &c.) are light crops, then soil becomes more rapidly exhausted by growing grain. 4. The soil may increase in productiveness and at the same time lose in fertility, due to the fact that the returns in the form of manure are more soluble and available as plant food than the same constituents would have been had they remained in their original insoluble condition. 5. Entirely apart from these considerations, it is not possible to save all the manure, and in practice one-half to two-thirds of the manure is wasted.

Farmers who raise large herds on small areas find that they can raise heavier crops than their neighbours who constantly raise grain, and they then rush to the conclusion that the fertility of their soil is being increased by stock-raising; but such a practice cannot be indefinitely sustained unless stock foods are purchased, which leads to a robbery of other farms. It is quite true that one farmer can increase the fertility of his farm at the expense of his neighbours, but the same results can be attained without stock-raising. Indeed, Britain is acting on this principle to-day; she scours nearly the whole earth for foods and fertilisers, thus robbing other countries, especially America, of their scientific wealth. The result of this process is, that Britain can sustain an increased population, but the population of the world cannot be increased. As a proof that stock foods are more exhaustive on the soil so far as nitrogen is concerned, the standard rations in my chapter on population may be compared, when it will be found that about 50 per cent. more protein per acre is abstracted from the soil in the stock ration than in the human ration; and even then it must be borne in mind that in the stock ration only the *digestible* protein has been reckoned. In order to make the comparison a fair one, the total protein being reckoned in both rations, the exhaustion would be over 80 per cent. instead of 50 per cent. more in stock rations than in those of human beings. However, if half of the stock manure is returned to the soil, while all the human excrements are wasted, that is another question.

I shall now, however, even take it for granted, for the sake of argument, that my conclusions on the manure question are all false, and yet neither the Malthusians nor the agricultural professors have the ghost of an argument to depend on. If they can increase the fertility of the soil by raising stock, I can do exactly the same thing by raising human beings. If they can save the manure, I can save the excrements. If they argue that the excrements are more filthy than the manure, then I admit that they are right if they have reference to economic men who eat meat and cook their food; but as my human animals can get no meat and must live at least as naturally as domestic animals, the manure and the excrements are on the same level in point of respectability. The excrements from economic men are also more valuable than the manure from domestic animals, for when foods are cooked or bolted, much of their nutritive properties and constituents is wasted. Thus we find that there is not one scientific argument in favour of stock-raising; on the contrary, there are objectionable features connected with every issue. Of course, it gives employment to many labourers; but, as we shall see, this is also very objectionable, and if I cannot find more respectable employment, I will admit the weakness of my system.

This discussion would not be complete without a word dealing more especially with the moral aspect of the question. The attempt to persuade farmers to raise stock merely as manure machines having proved futile—for they would prefer animals which could convert all their food into beef or milk, thus producing no manure—the agitation had to be kept up by other methods. The manure-machine theory was not a practical question amongst farmers who depended upon their beef and milk to pay the interest on their mortgages; and the science of soil-exhaustion was too technical a problem. The only means of escape was to drift into the hysterical method, just as the father of neo-Platonism solved the philosophic dualism. In America the spasmodic method of conciliating a dualism is called a "boom." History is full of such illustrations, some of which I have already pointed out, and many of our political economists, finding natural laws to

be more potent than their theories, have swooned away into
happiness.   If the sun's spots are the cause of commercial
crises, and if they cannot change these spots, why do they not
seek some other employment?   Reasoning from analogy, it
will now be quite natural for some agricultural professor to
suggest that our agricultural depressions are caused by the
man in the moon.   Stock-raising having assumed the essential
characteristics of a trade, it carries the idea of respectability,
and is thus practically no longer a branch of agriculture.   The
idea of being in some way connected with pedigrees tends to
elevate the social standing of the stockman, although he may
not be able to trace his own pedigree beyond his grandmother,
and the making of a voyage to England for the purpose of
purchasing a famous cow or hog, thus also getting his name
into print, places his position of respectability beyond all
question.   In order that these animals may be kept in a condition worthy of their noble blood and distinguished ancestry,
they must undergo a process of development, the end of which
is not yet quite reached, but may be illustrated by the life of
the Strasburg goose,[1] where 50,000 geese are fed annually for
the purpose of providing diseased livers for European gluttons.
The aim in feeding this goose is to produce as much fat as
possible, where it specially accumulates on the liver, producing fatty degeneration of that organ and the tissues of the
body ; and for this purpose the victim is closely confined in
a coop, or rather a small department thereof, where it has
scarcely room to move, the head projecting through an aper-
ture.   The coop, with its inmate, is then placed in a dark and
hot cellar.   The fattiest foods are given—such as maize soaked
in water ; charcoal and salt are put into the drink-water, and
the mess is squirted down the bird's throat.   In four or five
weeks the creature grows morbidly stout, and can hardly
breathe.   When the victim of this shocking cruelty is slaughtered, the liver weighs between one and two pounds, and
three to five pounds of fat may be roasted out of its body.
A visitor to any of our Christmas fat-stock shows cannot fail
to perceive that the Strasburg goose is the logical conclusion

---

[1] This goose has been known from Roman times, and is widely known
in France, where it is fed for *pâté de foie gras.*

of our live-stock industry; and the authorities may be congratulated on the rapidity with which they are approaching this ideal. The feeding of cows for milking competitions is approaching the same standard. The test of one cow for a period of seven days is made to decide the yearly milking capacity of the whole breed, and is used as an argument against other stock, pedigreed and unpedigreed. The object is not to arrive at any truth, but to hoist the price of the breed; and the herd-books are so manipulated as to limit the number of pure stock, thereby, in sympathy with the soundest principles of political economy, increasing wealth by making its commodities scarce; and many newspaper editors are induced to countenance and support this method of wealth-production. These animals are entitled to draw on a sufficient area of land for their support, which correspondingly reduces the area for millions of unpedigreed human beings. If any mortal creatures ever suffered the consequences of their ignorance and stupidity, it will be those who have attempted to establish pure breeds of dumb brutes and rank them between the upper and lower classes of human beings.

While we are staring at ourselves and at one another in boastful admiration of man's power over the dumb brutes—over their plasticity in his hands—what is taking place? The vital functions of the animals are being impaired, and the tendency to diseased obesity—an accumulation of effete material in the system caused by insufficient exercise and excessive gluttony—is regarded as a measure of the improvement of the breed. Even if these huge layers of fatty degeneration were fit for human consumption, they would be quite inconsistent with the orthodox theory that protein is the most valuable nutrient in human rations; and when such "baby-beef" is defended on the ground of the greater tenderness of its muscular tissue, the argument is only a proof of the degenerated condition of the consumer's vital powers. The milking qualities and the fecundity of the breed are diminished, disease follows, the breed falls into disrepute without compensation to the owners, and some hitherto obscure and unpedigreed variety is pushed to the front, only to undergo a similar career.

The intrigue, however, is good for trade, gives respectable

employment to a host of veterinary surgeons, Royal Commissioners, and other Government officials, and strengthens the powers of the politicians who are engaged in the discussion of measures for the prevention and spread of contagious diseases. The system is another comment on man's supernatural powers of objectivity—his mastery over the laws of nature. In his way of thinking, the animal which possesses the highest economic value can be made, by much cherishing, the fittest to survive.

Despite the uselessness of live-stock from every point of view, they are always classified in the category of our national wealth. By reckoning the enormous areas of land required for their support, the number of factories, farm buildings, hedges, fences, the quantity of labour, tools, machinery, and other fixtures, conveniences, comforts, and appendages, and, above all, their influence as wasters of soil-fertility, there remains but one conclusion, namely, that they are the greatest destroyers of wealth which a nation can sustain. Even setting aside the scientific view, stock-raising cannot be defended from the economic standpoint, for wealth could be equally increased by making scarce those other commodities required by the immensely increased population caused by the removal of domestic animals.

We thus find the same law operating in the extinction of pedigreed human beings and pedigreed domestic animals; but the case is different with reference to the unpedigreed varieties. In the former instance, the tendency to extinction is caused by drawing on too much land and labour, while in the latter instance there is a tendency to degeneracy and consequent extinction of the poorer classes of human beings caused by the lack of power or liberty to draw on sufficient land for their support; although, with reference to the unpedigreed varieties of domestic animals, they usually enjoy sufficient land for their support, neither too much nor too little, and are therefore the fittest to survive. In this respect they are substantially on the same footing as slaves. Chattel slavery is not inconsistent with the perpetuity of the enslaved race; on the contrary, the reverse is the case, for it is the tendency of masters to force their slaves to exercise their structures to the fullest tension without strain, the rations allowed being also conducive to the same end, and this is precisely the natural law of development. It is the interest of the slave-owners to preserve this nice medium, although, through ignorance, the requisite degree of perfection is seldom attained. Development is measured (1) by land area, and (2) by exercise of function, the former being the physical and the latter the moral aspect, and they are closely related to each other. It is not an error to include labour in the category of land, for the area required for a person's support cannot be increased or diminished without affecting labour, the only question to be decided being how far the areas are chargeable to those who draw on more land, or what returns they make for the absorption of their chosen quantities of land and labour—how far they are chargeable with the land required to support the horse which or the slave who works their parcel of land.

If this economic law of extinction were confined to domestic animals, the results, apart from the shocking cruelty of the method, would be more beneficial than otherwise; but unfortunately economic forces are also operating upon our agricultural plants. In our blind attempts to force them to survive after they have become weakened through excessive fostering, we find ourselves engaged in constant warfare against all kinds of weeds which, by virtue of our very efforts to exterminate them, become the fittest to survive, because their struggles for existence become so much the greater. Their seeds and roots lie exposed during the most unfavourable seasons, while the germs of our agricultural plants are comfortably housed; and like our live-stock, the higher the price the tenderer the care bestowed on them, thus causing the fostered varieties to run out. The plumpest grains, which are the weakest and least nutritious, are used for reproduction, and receive the highest prizes and awards at our agricultural exhibitions—just on the same principle as that adopted by the judges at exhibitions of domestic animals—. the basis of the judgment being the ability of the grains or animals to please the eye. It is not surprising that popular indignation is so often aroused over such abstract conceptions of justice, and that opinions differ as widely as those advanced respecting happiness or other abstract qualities. The theories which have been advanced with reference to the milling properties of different varieties of wheat have done more injury than the utter destruction of wheat and mills could possibly have done. And yet it is believed that the clamourings for justice at our agricultural exhibitions can be appeased by the appointment of one judge in each department instead of three. These exhibitions are not intended to benefit agriculture, but to draw crowds and to make money. They are purely speculative enterprises, and the methods of conducting them are purely economic. Another effect flowing from the diminished vitality of agricultural plants is the increase of all sorts of insects injurious to vegetation, whose havoc is the cause of intensifying alarm. The remedy for all these evils is supposed to be the establishment of agricultural colleges and experiment stations at the public expense. This method receives sanction because it encourages trade, the world being scoured for new varieties of plants and new breeds of stock, and respectable occupations are opened up for professors of agriculture, dairying, botany, entomology, veterinary science, &c.; and the increased demand for legislators strengthens the "powers that be." It is no figure of speech to say that the reason why we have to struggle so hard for a living is because it is our desire to obtain the most wealth with the least labour. Under the persistence of economic and objective rule, the coming race of men will be that which can best subsist on weeds; and when they, in their turn, are brought under objective control, being cultivated like the agricultural plants of the present day, they will run the same course and suffer the same destiny.

In England "grass is king," and we boast as much of our agricultural as our political monarchy. Least of all our agricultural plants is grass under artificial control, especially in permanent pastures, and this is the reason why he is king, and why all other plants are his subjects. The logical conclusion of this state of things is, that our isles will eventually be overrun with grass, while our domestic animals, being under economic rule, must disappear, and we shall then find ourselves in our original state of barbarism, living on the flesh of wild beasts, should any species survive by escaping domestication. But even we ourselves, as economic animals, cannot survive to witness this event, and if the whole world by that time becomes civilised, our race cannot escape utter extinction. If grass is to remain the king of agricultural plants, the consumers of grass should, one would think, become the king of animals. But if domestic animals must go, as science has pointed out, king grass must go too, must be dethroned, and the manner of his going has already been sufficiently indicated.

Agriculture is controlled by the same law which is bringing about the destruction of economicism—the idea of man's supernatural power over nature, intensified by the theory that vital laws are subordinate to those of the physical world. The former is a mistake of philosophy and religion; the latter a mistake of science. The increase of the world's food-supply during the past half-century is the most disastrous aspect of civilisation, and is due to three causes:—1. The discovery of the value and uses of artificial fertilisers. 2. The rapid expansion of cultivated areas and the enormous increase of live-stock on Western plains. 3. The improvement in agricultural machinery. Fertilisers increase the supply of available plant-food in the soil, and for some time may also increase the yield of crops; but, owing to the practice of making the phosphates soluble, the vitality of the plants becomes impaired, and the nitrogenous fertilisers cannot be perpetually substituted for organic matter, so that, in the long-run, these fertilisers are more injurious than beneficial. The use of animal manures, which tend to exhaust the soil more rapidly than grain-growing, is the result of another pernicious theory; and machinery only aids the other two forces in hastening the destruction of agriculture. So long as the soil's fertility is permitted to go to waste, there can be no remedy, and all our so-called agricultural improvements are a delusion. The reaction must come sooner or later, and then the sufferings caused by the pressure of population against the means of subsistence will be most intense.

K

CHAPTER VII.

THE THEORIES OF THE STATE.

THE weaknesses of the State may be classed under three headings:—1. The numerous theories which have been advanced with reference to its functions, and the results of human action, based upon these theories, which have been entailed. 2. The same economic ills prevail under all forms of government. 3. The State has been controlled, through its rulers, by economic and other abstract theorists.

It may be said that the weaknesses, contradictions, and absurdities of economic theories, which I have pointed out, are no arguments against the State, for it existed long before the age and reign of economicism. Very true; but it cannot be denied that economicism is an evolution of prior forms of objectivity. If the laws of nature had been man's first discovery instead of the theories of ideas, and obedience thereto cultivated, the State, so far as our present conceptions thereof are concerned, could never have received origin or development. No statute law, "by the grace of God," has ever been enacted except in conflict with some law of nature. The farther we remove ourselves from the laws of our being and development, and the more we sever ourselves from the means of our subsistence, the greater becomes the necessity for the existence of the State. If, to-day, the disciples of abstraction drew their sustenance from the same source as their supernatural ideas, there would be no social problem to be solved. If our foods were divided into two classes, the scientists getting the products of the earth and the abstractionists the manna from heaven, the social problem would be solved in forty days. It does not require half an eye to see that the greater the development of abstract theories, the more deplorable grows the social situation.

It is important to understand the various theories which have been advanced with reference to the origin of the State, for its scope and functions have emanated from these theories. According to some political philosophers, the family is the model of the State, although without consanguinary relationship. This is the Patriarchal idea, the supreme ruler being the father of the governed, and all conceptions of humanity must therefore be excluded. The tribe is an extended family, including several groups of blood-relations, and the chief may be the hereditary head of a single family. Patriarchism, of which the Jews and Chinese are instances, is repugnant to the Aryan idea; it tends to Imperial absolutism, blunts all noble aspirations, adheres too exclusively to the divinely ordained traditions and ceremonies, and allows pretext and scope for State interference in the most trivial affairs of the subject.

In antiquity and during the Middle Ages the idea prevailed that the State was a divine institution, founded, governed, and maintained by God, who administered the affairs more or less directly. The king governed the State as God ruled the world. The same idea prevails in modern times, as may be gathered from the founding of the United States of America in Washington's Inaugural Speech to Congress, 1789; and the Senate, the House of Representatives, and the State Legislatures are still opened with prayer. In the Jewish theocracy the work was direct and through divine revelation. That the State was indirectly founded and governed by God was the theory of the Greeks and Romans. But although their institutions were human, not theocratic, yet all public undertakings of importance were preceded, in ancient times, by prayer and sacrifice. The will of the Roman Gods was discovered through omens or auspices, indicated by the movements of birds, and the occurrence of thunder and lightning in certain portions of the sky. The Christian idea of the relation between Church and State is made plain by the words of Paul: "Let every soul be in subjection to the higher powers: for there is no power but God, and the powers that be are ordained of God" (Rom. xiii. 1). The authority of our earthly rulers is therefore directly politico-divine, unless Christ's injunction, "Render unto Cæsar the things that are Cæsar's, and unto God the things that are God's," has some modifying influence. The weakness of the theocratic theory, both the subjective and the objective phase, is the worship of earthly demigods, the palliation of all sorts of errors and abuses in the State, and the tendency to destroy individualism by absolutism. It divides society into masters and slaves, and encourages political philosophers to flatter despots.

The theory that "might is right" is attended with identical consequences, leading to the belief that there are no rights except those enjoyed by the strongest members of the State. This idea originated in a warlike age, and has special reference to physical strength ; but it is difficult to understand how it can be separated from the strength which the oppressed classes have acquired from time to time in the abolition of the coarsest phases of feudal iniquities. It is absurd to maintain that might acquired by the bullet is right, while might acquired by the ballot is wrong. Again, it is not necessarily correct to say, that the acquisition of this form of right is due to strength, for it may have its ground in the imbecility of those who are forced to abandon their rights. This principle is aptly illustrated by the French aristocracy and clergy, who, at the breaking out of the Revolution, were so weak and silly that they could but faintly grasp the idea of resistance. There is no law in nature by which right obtained by might can be perpetuated so long as the oppressed live more in harmony with the laws of their being and development than their oppressors. It does not alter the case to say, that the strongest do not lay claim to their rights; for the extinction of the enjoyers of rights leads also to the extinction of their rights. The might-right theory leads to a constant test of strength between the oppressors and the oppressed ; it is vicious and barbarous in the extreme, and no hope for a solution of the social problem can be entertained so long as it sways the public mind.

Then, again, we have the theory of contract or social compact, which operated fatally in the French Revolution. In this instance the individual surrenders his natural rights to the State in return for security. This is said to be the transition from the state of nature to the social state. Any form of evolution from the state of nature, directed in any channel whatever, could only have originated in the abstract theory that man is a supernatural animal, and is therefore not obliged to obey the laws of nature. The more this theory is developed, the sooner is the day of its doom. If the security of property is meant, then we are confronted with the question relating to the right to labour and accumulate property; and if the security of person is meant, then I want to know what incentive there would be for aggression against the person except when it bears some relation to his property. In the discussion of this question, it is necessary to have a precise understanding as to whether the State is ordained of God or of the people—whether it emanates from the dead or the living. What are, also, the powers and functions of the prospective living—the generations yet unborn? Naturally, the theory of social compact leads to all sorts of speculations as to breaches of contract on the part of the State, and so justifies revolution and anarchy. It tends to a condition which renders the existence of the State impossible, if not a pure fiction. A social compact implies equality amongst the individuals forming it, and it is the condition of inequality that necessitates the existence of the State. That all men are born equal, as declared in the American Constitution, has no practical import unless they remain equal after they are born; and if all remain equal, socially and politically, who is to rule, and who are to be the ruled? By driving this theory to its logical conclusion, the masses have a grand opportunity for asserting their equality, and of overthrowing all despotic and theocratic oligarchies. The history of France during the past century amply proves that the abstract conception of "liberty, equality, and fraternity" is a failure, and produces the direct reverse of the results intended. The French Revolution is the source of the armed despotisms in Europe at the present day. Is that liberty? Is it equality? Is it fraternity? Is the American nation, whose Constitution is founded on the same basis, a model of anything that is conducive to the well-being of humanity?

The illusion that no State can arise without the mediation of some prophetic spirit, such as Moses or Mahomet—demigods who descend with missions and titles from heaven—is naturally courted by aristocracies, for they are aids to oppression and the development of absolute sovereignty. In democracies, the same illusion illustrates how it is possible for a political function to remain for a time after the structure is gone; for the reverential awe which the State inspires—the belief in the supernatural powers inherent in the objects—remains after the crown, the wig, and the woolsack are removed.

In other theorisings, the speculation is rife that the State had its genesis in the social impulses inherent in human nature; in man's political consciousness—in the theory that man is a political animal. This is a direct thrust against the theory that man is naturally a war animal, and requires the State to maintain peace, to protect his liberty and his property, and to give him security in the pursuit of his happiness. In this case, the State must, one would think, be indirectly a divine institution, for God must have implanted these politico-social "instincts" in our nature. The argument is weak from the fact that, like religion, the instinct is not universal in man; for there are individuals and political parties whose instinct it is to overthrow the State, no odds what is to reign in its stead. It quashes the widespread theory that the State is a necessary evil, and it makes it a necessary good. It can only lead to the same conclusion as the theocratic theory, namely, the "realisation of God's kingdom upon earth." All instincts must be universal in human nature; otherwise the attribute must have originated in an artificial manner—must have been perpetuated by habit, and is therefore not instinctive.

The solution of these theories, however, would be of little or no avail, for there are numerous other theories with reference to the end of the State, which do not necessarily bear any relation to its origin; and there are, moreover, theorists who maintain that the State has no end at all, but is only a means through which individuals can attain their ends. The ancient theory, which destroyed all individualism, was carried out as if the State only had an end, to which the individual contributed his part. Everything was sacrificed to the welfare of the State. In modern methods of speculative thinking the State is merely the means through which the welfare and happiness of the individual can be most efficiently secured—at least, the welfare and happiness of the greatest number. The weakness of this theory is, that the champions of this philosophy have not pointed out what the end of the individual is which the State is called upon to forward; neither have they been able to define the meaning of welfare and happiness. In the absence of these definitions, it is impossible to define what the State is, and the whole fabric becomes a myth. If the people form the State, then all must help each, although each may be at liberty to wage war against all. Is the Deity the source of sovereignty, then He is invoked to help each, although each may enjoy the liberty of defying the divine authority. Any State that can exist under the pressure of such an outrageous dualism must be more than supernatural.

Following the philosophers of the German illumination, we find the theory prevailing, which has also gained a firm footing among Individualists in England, that the true functions of the State are restricted to the assurance of individual rights; and Kant expressly rejected the theory of the welfare or happiness of the individual. Fichte expresses the same views, and Humboldt restricts the State sovereignty to "the maintenance of security against both external enemies and internal dissensions." Mill regards government action as being less effective than voluntary action, each person being the best judge of his own interests; but he would permit the State to interfere, for example, when the consumer cannot be capable of judging the quality of goods, and would allow the State to educate children and even adults, and regulate the hours of labour. His rule of State non-interference has so many important exceptions that it may be regarded as a vague theory of public expediency. Herbert Spencer, who regards the State as an aggressor, not a protector, when it goes beyond the protection of life, liberty, and property, proclaims every citizen as having equal rights to pursue his happiness and enjoy the fullest exercise of all his functions, limited only by recognition of the same rights in others. Humboldt excludes education, religion, and morals from the supervision of the State. Hegel, agreeing with Plato, contends that the function of the State is morality, and the realisation of the moral law. What is morality? Whence arises the moral law? The public welfare, according to the Romans, was the real end of the State. One of the latest theories is the development of national capacities and a national spirit in directions consistent with human destiny. What is the destiny of our race? Austin finds it convenient to regard the State as being moulded by the will of a dominant body.

An important principle in relation to the State is the franchise. Is it a right belonging to the subject, or a privilege granted by the sovereign? It is evident, in view of the confusion of theories just enumerated, that this question cannot be answered through any abstract medium. Who is in actuality the sovereign? To say that the king is sovereign does not make him so, unless it can be shown that there are no powers behind the throne. History proves that the owners of the land are the sovereigns of the earth, and if humanity owns the land, it is also the supreme sovereign, whether we recognise the fact or not. When we once recognise and act upon the fact that humanity[1] is the supreme ruler, the social problem will be solved; and the source of the whole social struggle reposes in our coronation of false sovereigns. Unless we stop posterity, we may as well give up the struggle; for, especially when ancestor-worship is dissolved, we cannot force it to discharge our obligations when we leave behind neither the raw material nor the strength to appropriate it even for its own subsistence. The weaker we leave posterity, the greater grows its sovereign strength, and the more peremptorily we shall be called upon to render an account of our stewardship. The owner of the land is the owner of the crown—the absolute owner, who reigns and

---

[1] There is a sharp line of demarcation between the humanity of the past and that of the future, which will be drawn in a later chapter. In the sense here indicated posterity plays the most active part, and we may regard ourselves as the posterity of our ancestors.

governs, and from whom we derive and enjoy the franchise. Jovism and Jahvism, as well as those other so-called sovereigns, reason and justice, must relinquish their sovereignty, and the nation, even though it be guarded by its language, its laws, its literature, its religion, its ranks, and inherited sympathies, must become the footstool of the absolute.

This scientific fact is supported by the history of three thousand years ; and the constant swerving from Jovism to Jahvism, from monarchy to democracy, and *vice versâ*, proves the hopelessness of the struggle. The most ancient form of government was the theocracy, the rule of priests who, if not always Gods themselves, derived their sovereignty more or less directly from the God or the Gods. In ancient Egypt the original rule was directly from the Gods, and in later times the kings, though human, were descendants of the Gods, their authority being circumscribed by the divine law. In the Jewish theocracy, under the Mosaic dispensation, Jahveh himself was king—legislator and ruler—to whom all the soil of the promised land belonged, the various families being mere tenants who had to pay rent in the form of a tenth of the produce of the land and flocks for the maintenance of the priests. This form of tenancy tended to social equality, no Jew being a slave, and in the year of jubilee a fresh division of the soil was made amongst the tenants—not amongst the king's elect, as in mediæval and modern times—after the jubilations on war's bloody fields. But this form of sovereignty, in all ancient States, failed ; and even the Jews demanded a human king in order that they might be like other nations, divine rule having proved to be a miscarriage of justice. The deities themselves, which the nations created in their own image, failed to recognise the legitimate sovereign of our earth. Then came the transition to the mixed State, a confusion of theocracy and monarchy, whose days are also numbered, and to-day monarchies and democracies, as well as all other forms of human government, are crumbling to the dust. The fallacy that virtue can be created out of two or more forms of vice has not yet been cognised. The French Revolution, which I regard as the highest type of all history, is the strongest proof of the fact that we have not

yet recognised the true sovereign of our earth, despite the
abstract conceptions of the Rights of Man which gave it birth.
Moreover, the lord paramount of the soil, who sways the
political sceptre, must also wear the sacerdotal stole. The
virtual collapse of feudalism, which weakened the positive
powers of the lords and gave inception to the armed despotisms of the present day, has fixed no proneness towards the
cause of humanity; indeed, the reverse being the case, the
only legitimate conclusion is, that the development of abstract
methods is the most disastrous of all human failures.

But my remarks have been essentially confined to the
historic State. It takes many years for facts to become
matters of general history, so that the economic State still
remains, for the most part, incidents of current literature
and observation. In drawing a faithful picture of the economic State, it is not necessary that the artist should confine
his sketches to any existing form of government, for all political institutions in civilised countries belong to the same
breed, despite the numerous varieties of aristocracy, democracy, and theocracy. Feudalism has developed in politicism
—politico-economicism—and the statesman has evolved into
the politician.

The politician has virtually ceased to coerce his victims for
their spiritual good, and has confined his energies, his ambition, his patriotism, his brilliant abilities, his modest refinement, and his religious sentiments to their coercion for their
material or economic good. Barring the author, the critic,
and the journalist, the politician is the only man who requires
no learning. Politicism belongs to the "innate moral sense."
Like the poet, the politician is born, not made. This is the
reason why there is no ultimate dualism in politics; for all
politicians are unanimous in the conviction that they constitute the State, that society should work for the State's welfare, and that they have "vested rights" of a social and
political character—both for themselves and their "heirs and
assigns for ever." It is of no special consequence whether they
own the soil itself or its products, including the labourers, for
the land is of no use save for its products, and both forms of
holding—*i.e.*, the land and its products—are now accompanied

by equal increments of respectability. In this manner, the dualism of land and capital has been solved—also the dualism of Church and State—and the rights of the persons in this trinity are equally inalienable, their rights to rank and to happiness, as well as their material rights. They enjoy implicit faith in the theory that nature is governed by their laws—that our knowledge of the Absolute is embodied in their statutory enactments. They are divine by sentiment, behaviour, privilege, and power, if not also by pedigree and descent. They create wealth in a respectable manner, by consuming the products of labour, thereby making commodities scarce and dear, and furnishing employment for the "proletariat." In aristocracies, their generosity is exercised by working without wages, merely demanding in return increased rents, profits, and briefs; in democracies, monopoly and wages are the cost of legislation. There appears to be a trifling dualism here, but it would be uncharitable to dwell on trifles so long as there is so much unity and harmony of thought in the grand doctrines and ethics of politics. Besides, all minor appearances of inconsistency are bridged over by private donations to charitable institutions, and by a free-handed circulation of money during election campaigns, when the sympathies of our politicians for the masses are most keenly aroused, and when pledges are most rife—promises of law for the poor and justice for the rich. Pledges for the extension of the franchise is a common form of generosity bestowed in return for election or re-election.

The economic State is divided into two departments extraordinary, namely: (1) The Society for the Production of Wealth (without labour), and (2) The National League for the Protection of Property. These, as well as all the sub-departments, are run by syndicates, rings, "bosses," caucuses, conventions, &c. It is an imperative duty of the functionaries to see that hereditary sinecures are created or augmented for the purpose of furnishing respectable employment for needy partisans who have no visible means of subsistence. Should malcontents arise in any quarter, especially amongst those who enjoy the franchise, or should a depression in trade ensue in sympathy with the malignant spots on the sun, a Royal Commission is appointed to investigate the cause, which creates respectable employment for men of birth, education, and refinement, and then the evidence is most circumspectly shelved to receive the most prayerful attention of the party—after the next election. When the theories of our political economists are threatened with discredit or extinction, due to trade depression, a Royal Commission is in demand for the purpose of gathering more statistics in order that our economic thinkers may subdivide their subject, it being much too broad for the wits of any one man in a single lifetime. Should there be an agricultural crisis, then our live-stock speculators, who are organised to represent the whole farming community, are examined to show cause, and in the evidence it appears that more pedigreed stock is wanted to maintain the fertility of the soil, that the pedigrees must be purer and less spurious in order to make them scarce and thus enhance prices, that State funds are required to examine the herd-books, and, in order to prevent a collapse in prices, the reports of the prevalence of contagious diseases amongst blooded stock must be suppressed. Then arises an alarm lest people should catch trachina, tuberculosis, or other contagious diseases which infest domestic animals; and then more doctors and inspectors are required to diminish the high death-rate due to over-population, and to prevent food adulterations, which have been increased for the good of trade. Another Commission is in request to inquire into the labour grievances—to devise ways and means to enhance wages and shorten the hours of labour without interfering with the interests of the capitalists; but the agricultural labourers and tenants, being unorganised, or their votes being secure or non-existent, fail to give employment to politicians in the Royal Commission line of business; they are therefore non-producers, because they do not encourage literature or trade; and for the reason that they keep "scrub" stock they also fail to encourage veterinary science. They take no part in the restoration of business confidence, without which trade languishes and wealth becomes demoralised. Should destitution become so great, owing to the mass of unemployed labourers, that the State becomes endangered by the spread of Socialism, then the surplus wages must be forced into the State treasury to provide against old age or infirmity, thus compelling the malcontents to become partners in the public business, which coerces them to patriotism and strengthens the "powers that be," for the overthrow of the State would then involve the ruin of the lower as well as the upper classes. This sagacious stratagem has the effect of quelling Socialism by the adoption of the Socialistic principle, just as landlordism is abolished by increasing the number of landlords. Political events naturally become a respectable topic of conversation amongst the learned, who, although they may not have political aspirations themselves, cannot retain social respect without adopting a political cause. There must be a struggle against innovations in our universities and ecclesiastical institutions; for changes herein would be an acknowledgment of their imperfections, and would lead to the insinuation that Conservatism has a Reform aspect. Then, again, who has the brazen effrontery to deny that "vested rights" exist in literary, economic, philosophic, and theologic professorships, as well as in those departments of human affairs which minister to our material well-being? The title-deeds thereto are surely as sacred and inviolable as land tenures, or bonds and mortgages. If the theories of Adam Smith, Henry Carey, Malthus, the Church fathers, or those of the philosophers were extinguished, the "counsel for creeds" (as Professor Huxley puts it) would lose their means of subsistence just on the same principle as landlords or bondholders. If biology instead of chemistry were made the basis of dietetics, the occupation of our doctors would be gone, for the legitimate conclusion would be that drug-shops are a humbug. If wealth were based on plenty instead of scarcity, our lawyers would have to take angels' wings and fly, followed by our whole army of politicians—fly to the regions whence the land titles have been obtained. It would be a social and political crime to insist that the future demands a portion of our attention, for there are so many venerable ideas and institutions in the past which claim our undivided energies. Besides, a peep into the future would stimulate a proneness in the cause of humanity, and ruthlessly drag us out of the rut of conventionality. If our landlords, pursuant to their feudal contract, still did the fighting, a dreadful calamity might accrue, there being nobody to own the land or consume the luxuries obtained therefrom, and then there might be no title-deeds for lawyers to trace, so that two respectable sources of employment would be gone, and then there might not be sufficient wit or forensic eloquence to defend our tithes. All forces which tend to weaken human desire are destructive of wealth ; and as the incentive to crime becomes proportionately weak, economists and lawyers are brethren in the most significant sense of the word ; the clergy belong to the same fraternity, and all their interests are bound up with those of the State. A strong government becomes a legal, as well as a clerical and a political, necessity. When the regal authority becomes weak, whether through upheavals due to widespread distress and popular discontent or through licentiousness in the higher ranks, the State is strengthened by clamours and declarations of war. The demand for fortifications and other public works gives employment, is expected to awaken patriotic feelings, and to restore contentment and business confidence. It is easier and more pleasant to wage war than to fight, and a policy which tends to divide the · political power amongst the fighting classes becomes unpopular at headquarters, unless an undue prolongation of peace tends to desinecure military aspirants. State officials, being engaged in respectable employments, become the leaders of society, create wealth by encouraging desire and trade, thus furnishing employment for the poor and needy, and they carry civilisation into localities which would otherwise remain in the most degrading state of barbarism. Imitating the higher government functionaries, they become leaders of Fashionism, and their refined behaviour wins for them that degree of respect and adoration due to their social rank and political calling.

There was a time when our sovereign exercised a power in State affairs. For example, Edward VI. called the lower classes of his subjects "spittle and filth ;" but the sovereign of the nineteenth century has lost all power in this respect. Perhaps it would be more correct to say that he voluntarily resigned such authority, for its continuance would be a biting satire on the very civilisation of his own begetting. His authority is now limited to his personal influence over the ministers of the crown—in other words, it is identical with that of any subject in the kingdom. The power lost on the one hand, however, is gained on the other, leaving the sum-total unchanged. His power as master of economic ceremonies has vastly increased, and the greater the increase the greater the area of land required for the sustenance and development of court ceremonies. He is the most potent instrument in the quickening of our desires, in the encouragement of trade, and in the furnishing of employment for his starving subjects, not only on his own behalf, but also on behalf of all other court dignitaries who confer the same blessings under royal auspices. But in ancient times the king was also judge, priest, and general, all of which offices were divine. The general and the judge have lost their divinity, that of the priest is all but gone, leaving the "defender of the faith" alone in possession of the divine functions. But he is now the high priest of economic rites and solemnities; therefore economicism is divine—the faith of which His Majesty is defender.

Meanwhile, however, our State functionaries are so busily engaged in promoting the cause of civilisation that they have no time to devote to the material interests of the people. State ships, superintended by State inspectors, sink into the sea; State railways run into embankments, houses cave in, boilers explode, and human souls are ushered into eternity. Their predestinated time has come; the will of God is done. Besides, the country is over-populated, and, the inspectors being saved, the Almighty has taken this method of enabling the fittest to survive, there being no chance for those born and bred in the lower ranks of society. But it is also necessary that the population should be multiplied and replenished in order to increase the wealth of the nation; for rent would otherwise be at a standstill if increased demands were not made for access to the means of subsistence; and as for houses, they could not increase in number and swell into huge and flourishing cities, save by virtue of an augmented population, and our manufacturers and traders would have no employment in the furnishing of such houses. Therefore, a population having intense and manifold desires must be encouraged in order to increase rents and enliven business. People must therefore be used as machines for the augmentation of national wealth, just as our children, under the Education Act, are converted into machines for rewarding teachers according to results. The government method of increasing the population is by means of State-supported illegitimate children and foundlings, and by threatening bachelors with increased burdens of taxation. Another method of remedying over-population is in the appointment of delinquent adulteration inspectors, who permit the population to be prematurely shuffled off by the sure, though slow, process of poisoning.

Meanwhile, also, private enterprise straddles streams with bridges, binds oceans together with belts of thought, soul speaking to soul across the deep in defiance of time and distance, lays railways and telegraphs, almost greases the North Pole, expands commerce, Christianises the barbarian and brings him under the yoke of civilisation, codifies our laws, gathers statistics, sends paupers beyond the seas, establishes charitable and educational institutions, founds and maintains hospitals, museums, libraries, asylums, &c., exposes adulterators, agitates against corn-laws, carries home-rules, conducts exhibitions, fat-stock shows, and experiment stations—in opposition to State machinery.

This chapter is a brief abstract from my certificate of character to the economic State, which I most respectfully dedicate to Socialism. This is the State whose powers are to be enlarged so as to control every sphere of human activity. It is not surprising that Individualism rises indignantly to revolt against such an organisation, and that Georgism shrinks from the idea of going past the half-way house. If the Socialistic plan, namely, the more rapid development of economicism, is effectually carried out, the return to Individualism will be sharp and through no circuitous route. Equally futile is it to hope for relief through continuance of Individualistic rule, or any modifications of its forms, and therefore no compromise between the two evils can be of any avail. So long as we confine our operations within the abstract category, and remain blind as to the existence of ulterior dualisms, the Socialistico-Individualistic phase of sociology will not enable us to advance one step towards the solution of the social problem.

A brief analysis of Socialism and Georgism in their relation to the State is here in order. The Georgic school begin and end their present programme with land nationalisation. The Socialists are also in favour of nationalising the land, but their method of appropriation is different. Henry George taxes the land to its full value, and deposits the money in the public exchequer, while the Socialists issue a decree declaring the land to be public property. The only difference between the two programmes is, that the former is evolutionary and the latter revolutionary, so that the Georgic method is more scientific. In both cases the land passes into the hands of the State, and the rent must, theoretically at least, accrue for the benefit of the people. In the ulterior effect there is, therefore, no dualism between Henry George and the Socialists. The latter, however, do not stop short at land nationalisation, holding that this half-hearted Socialism will have little or no effect in ameliorating the condition of the masses. This is quite natural ; for it is the tendency of all parties to confine their gaze to immediate results. The Socialists, for example, finding that they prosper under conditions which make labour plentiful and wages high, regard the nationalisation of capital to be the basis of prosperity. Agricultural labourers and tenants, finding that high rents are a bar to their progress, naturally sympathise with any movement tending to free the land from oppressive burdens and making it accessible. Their idea of prosperity is to have land cheap and its produce dear, while the Socialistic idea is to have agricultural products cheap and free access to capital. Some independent theorists, who do not derive their subsistence directly from land or manual labour, contend that high prices all round are conducive to prosperity, while others base it upon the cheapness of all commodities. In Georgic philosophy, we hear that if we own the cow we also own the milk, which seems to solve the Georgo-Socialistic dualism, Henry George having become an extreme

Socialist. The failure to recognise the fact that the owner of
the land also owns its products arises from the lack of aptitude
to draw a distinction between the products already taken from
the land and those taken after the land becomes nationalised.
A very large percentage of commodities is more or less
perishable, and require to be replaced by fresh draughts upon
land, so that the question to be decided is the rights of the
individual to labour upon State-owned lands for the purpose
of appropriating the products for his own use and that of his
"heirs and assigns for ever." Food, clothing, and houses
being perishable commodities, the State must eventually have
complete control over the individual, and may even prevent
him from building houses or erecting machinery on land. It
is idle to seek any ultimate difference between Georgism and
Socialism; and we are therefore confronted with two conditions, namely, (1) Evolutionary Socialism, and (2) Revolutionary Socialism. The revolutionary party is damaging its
own cause by attempting to get all it can as fast as it gains
strength to procure it, while under the evolutionary system
future results are to some extent matters of reason and calculation, not of prejudice and passion. There are many well-
meaning people who would join the Socialistic movement if
they could foreshadow the results of their actions, but so long
as the means of subsistence remain precarious, they will
retain grasp of their property until it is forced out of their
hands. The question is not who is entitled to the property,
or who has the greatest power to acquire and retain it; for
even if the propertied classes suddenly released their titles,
it would be unwise on the part of the revolutionists to attempt
peremptory possession. All safe institutions must grow. To
dethrone a king and elect a president in his stead does not make
a republic. We must grow from bad to better, as well as from
good to worse. The real dualism in Evolutio-Revolutionary
Socialism is whether the process of reform should begin by
nationalising (1) land, or (2) capital; but as I am not working in the abstract category, I do not feel disposed to give
advice on this question. All I can say is, that all scientific methods are evolutionary, and if the nationalisation of
capital can be accomplished by evolutionary processes, that is

not a matter for me to discuss. With reference to the relation between Socialism and Individualism, it is an inconsistent coincidence that the father of Evolutionary Socialism, Herbert Spencer, is the champion of Individualism. In his work on "Social Statics" Mr. Spencer advocates land nationalisation, and he does not appear to have predicted the extreme Socialistic consequences of his own doctrines. None of the Socialist philosophers appear to have discovered the absurdity of their theory that every man can be entitled to the products of his labour. In the first place, no man can get the whole produce of his labour so long as there are politicians, lawyers, doctors, economists, philosophers, priests, speculators, weaklings, and other non-producers to be supported from the toil of the labourers. It is the rapid increase of the non-producing classes, especially of officialism, that is the cause of nearly all our woes—people who procure respectable employment and draw on large areas of land for the purpose of maintaining useless ranks and ceremonies. Secondly, the theory leads to the presumption that every individual has the right to labour and accumulate, for himself and his "heirs and assigns for ever," all that he is able to produce over his current necessities—that is to say, he is permitted not only to work his own plot of land (or receive an equivalent), but also to waste or hoard the substance of some other person or persons. The hoarding of valuables is one of the greatest evils in our social system, and so long as the practice exists an unequal distribution will be the inevitable result so long as there exist designing men. Besides, if everybody received the full produce of his toil, there would be no use in attempting to restrict the hours of labour. People would be more apt to work sixteen hours a day than eight, as many business men do now, and then the quantity of commodities hoarded would be proportionately increased. The monopoly of land products is just as iniquitous as the monopoly of the land itself—indeed, the product is a portion of the land—and the greater the quantity removed the less the quantity in reserve. The theory that everybody has unlimited right to labour must be abandoned, for it implies a corresponding right to monopolise, . to waste, to hoard, or to destroy. Not only so, but it also

confers a right to the individual to bequeath a life of idleness
to his heirs, and our social condition would then be the same
as it is now.   An equal right to the land must carry an equal
right to its products.   By what process of alchemy can rights
be maintained under a system in which the poor man's franchise has the same political power as that of the property
owner?   Equality of property can only be kept separate from
equality of voting power so long as the cunning of the privileged classes can be exercised in keeping the masses ignorant
of their rights, or in creating splits in their ranks.   And yet,
if the incentive to accumulate property be taken away, the
main stimulus to labour, in economic men, would be greatly
weakened, if not wholly suppressed; and it would be unnatural to entertain the hope that the State, through its
officials, could create a wholesome impulse in the individual.
Socialism can only enhance wages by making commodities
scarce, and for this purpose there must be, as under our
present regime, a great number of small men with great desires—not of great men with no desires, which is required
for the maximum of scientific wealth—so that instead of
having five or six millions of big gluttons, as at present,
everybody must become a little glutton in order to furnish
abundance of labour and high wages.   This principle is paralleled by the abolition of big landlordism through the creation
of numerous little landlords.   With all its defects, Socialism
will have this advantage, namely, that its leaders, should
they get into power, will use their utmost endeavours to
make their system work smoothly; the evils may not be
seriously felt until their successors in office commence to turn
society into a machine for their own aggrandisement.   Under
the same conditions history will always repeat itself.   I refer
to Evolutionary Socialism, although a desperate extreme of
Individualism will be required to force Socialistic ideas upon
the Teutonic character.   The Socialism of ancient nations
did not preserve them from extinction, and the same powers
of objectivity will cause the dismemberment of all modern
forms.

I need not dwell on the fact that the same economic conditions prevail in every civilised country, for such has not

been disputed; and with reference to the effect of economic theories upon the minds of our rulers, I have made the subject plain in my chapter on political economy. That the modern State is an economic institution is beyond all question. Efforts have also been made to help the cause of agriculture, but, as I have shown, this is also completely under economic control. Our Minister or Board of Agriculture cannot help the cause, but may aid in leading our agricultural interests to destruction. There is no use in wasting time to discuss the relative merits of Liberalism or Radicalism and Conservatism, or to enlarge upon such commercial strifes as Free-trade, Fair-trade, and Protection, or upon bounty-fed commodities. If commerce itself, as the basis of national prosperity, has proved a failure, it makes little difference whether trade be free, fair, or protected. Our Liberals and Radicals do not promise us any reforms which have not been a failure in other countries, notably France and America. Constitutional monarchies, despotisms, oligarchies, and democracies have all failed; and in political history there is nothing better known than the constant tendency to swerve from one extreme to another. In structure the political State of modern history is a mixture of the Græco-Roman and the Teutonic, but the functions are so inconsistent with these ideals that dissolution, sooner or later, is inevitable. Just as with the animal body, the State structure cannot long remain when the function is gone; and the same law applies to professional life. When the State has no faith to be defended, the king cannot remain defender of the faith except as an abstract idea; and even the priestly stole emaciates into a rag. When no disease is there the doctor is nowhere. When divine rights and titles vanish, and when wealth ceases to be based on scarcity, the lawyer's occupation is gone.

In principle there is nothing in the structure of the modern State which has not been pointed out by Aristotle; and his "Polities" has braved two thousand years the battle and the breeze. He makes the following classification :—1. The good government of one person, Monarchy, when demoralised, forms a Tyranny (Despotism). 2. The good government of a few persons, Aristocracy, when perverted, forms an Oligarchy. 3.

The good government in the hands of many persons, Commonwealth, when depraved, forms a Democracy. These divisions, modified by economic forces, hold good at the present day, and the "Polities" of this great scientist and philosopher has swayed the minds of political thinkers in mediæval and modern times. Now, Aristotle was a heathen, and the question arises, How could the divine institutions of the Middle Ages emanate from the brain of a heathen philosopher? If it be denied that Aristotle was a heathen, two proofs, I think, will be sufficient to settle the question, namely, (1) he was a believer in false Gods, and (2) he denounced economicism.

As a system of education, the vicious tendency of politicism cannot be too firmly fixed upon our minds. It permeates the press, the club, and all social organisations to such an extent as renders scientific inquiry very dilatory; it tends to deaden the spirit of truth, and to arouse our most ignoble passions; it sharpens our appetites for sensations and political scandals, blunts our moral feelings as to the squalor and destitution which constantly rage all around us, and demoralises our whole character. The youth, having completed his education in the State school, now becomes a pupil of the editor who is likely to be a tool of the party, and whose livelihood depends upon the amount of poisoned breath which he can exhale into the languishing body-politic.

I have no sympathy with those speculative thinkers who maintain that the State, like all organic structures, is naturally doomed to decay. Nations, it is true, will continue to fall so long as the germs of decay continue to gnaw the vitals of civilised races; but let the wheel of fortune be once turned the other way, let every generation excel the strength and virtues of its predecessor, let the rule of the true sovereign be once recognised, let "liberty, equality, and fraternity," in their true meaning, prevail, and the national life be as imperishable as humanity itself, constantly gaining in vitality, instead of becoming tainted with decay and sinking into utter dissolution.

The universally prevailing superstition that the State can do something has its origin in its supposed divinity. It is the great antiquity of the institution which leads us to believe that it enjoys superhuman powers; and it is even still the universal belief that the State can solve the social problem. The people build the nation, and the State is the author of its destruction. The State is a consumer and waster of commodities, not a producer and sparer. During the whole rule of economicism the State has been attempting to build up trade, and this is the source of those commercial crises for which economic thinkers have blamed the spots on the sun. It has also attempted to build up agriculture, but similar results cannot be escaped. Our religion is a sad commentary on the theory that our inner consciousness can be moulded and developed by State authority. Granted that trade is a good thing, and that the State has at times been the instrument of its enquickenment, then what do we find? Who has not heard of complaints of trade depressions in many branches due to elections and political campaigns? Many business men are waiting to see which party will enjoy the reins of power, or what will turn up after the elections. One party pledges itself to rob one branch of business or industry by taxation, and the other party promises to rob other branches, and it is necessary for the business man to know whether he will belong to the robbers or the robbed. So it is with wages : one party agrees to rob the capitalist in order that the labourer may receive justice; another party reverses this order; and the independents pledge themselves to make everybody prosperous without robbing anybody. This is the political method of encouraging trade; and it must be the only legitimate scheme, because the State is a divine institution, and therefore cannot err. If the ruler monopolises all the divine attributes, then witness the business sensation, especially on 'Change, when he delivers a public speech or issues a political decree. It is significant that a lottery corporation operates on political principles, and yet some of us can see that it can only influence the distribution of wealth and check its production. In a similar manner, the State can only tax one portion of the community and relieve another portion, which, however, can only produce the desired effect until the equilibrium of economic forces is restored. So long as this power remains there can exist no such thing as "vested right" in property, for the State can arrange the methods of taxation in such a way that commodities are equally distributed in sympathy with the voting power of the people, should the franchise receive the widest extension. If political "might is right," then equal voting power carries an equal share in the commodities within the control of the State. Our social troubles are caused largely by the ignorance of this power which prevails amongst the masses. However unjust this form of right may appear, it is not inconsistent in its practical effects with any law in nature. The State has also temporary power to grant monopolies, which result in the pauperisation of the masses, and it can tax the monopolists up to the full value of their monopoly—that is, the function of the economic State is to grant and destroy monopolies; and its administration of justice consists in the punishment of individuals who, impelled by necessity or other economic forces, attempt to weaken or destroy monopoly. Herein resides also the power of the State; the greater the monopolies granted, the greater its power, for it can control the monopolists by pledging the continuance of the monopoly, and can control the anti-monopolists through pledges for its destruction. In the ultimate analysis, there can therefore only be two political parties, namely, (1) the defenders of monopoly (political objectivity), and (2) the monopoly destructionists (political subjectivity); and to this source may be traced the duplicity of our politicians. Even as a matter of mere economic justice, it has not been explained why those who make the monopolies valuable should be deprived of the benefits. The indirect scheme of taxation is one of the most potent instruments in the maintenance of monopoly. The indirect form induces us to believe that we pay no taxes unless we feel the hand of the tax-gatherer sink into our pockets, while in reality the more indirect the taxation the greater is the sum we have to pay —and the less we think we pay. If all middlemen engaged in the collection of our taxes were abolished monopoly would lose its props, and great would be the fall thereof. The State, as a lottery corporation, is the prince of humbugs on the ground that every citizen is compelled to support it, although, unless he enjoys the franchise, he has no voice in its destruction. When the winner receives his prize, he is not secure until its value is restored by taxation ; but so long as he is able to retain it, he has a title-deed to larger areas of land wherethrough he may maintain himself in the criminal condition of unbridled licentiousness and gluttony, enabling him to dispense with the exercise of his natural functions, to weaken the functions of the holders of unlucky tickets through overstraining labour, to demoralise his associates by their aping his desires, and to bequeath to posterity. his impaired structures and functions. All this, however, is good for trade, and that settles the matter, for by increasing the totality of desire wealth is increased in substantially the same ratio.

The abstract character of the State is now made plain from various standpoints, and all hope of attaining unity of thought and harmony of feeling through political methods may now be abandoned. The State is an economic institution, and our dualism therefore remains unchanged, namely, science *versus* economicism (scio-subjectivity and economico-objectivity).

CHAPTER VIII.

STANDARDS OF VALUE.

It will have been observed that I scrupulously evaded all allusion to money as a factor in connection with my inquiry; for had I done so I would have committed the tragic outrage of insinuating that it had something to do with the solution of the social problem. However, as the gold superstition is so firmly rooted in the economic mind, and as standards of value play an important part in my inquiry, I have found it necessary to devote a special chapter to the subject. As already stated, our economic thinkers strive to found our monetary matters on the soundest principles of political economy; and as these principles are fundamentally false, it is evident that their efforts will be barren of practical results. We have seen that the basis of economic wealth is an abstract conception—and it makes no difference whether it is called desire, scarcity, or happiness—and as wealth is measured by money, therefore money must be the measure of scarcity (if such can be measured), not of abundance. When commodities are so abundant that they bring no money, there is no economic standard for their measurement. To use a concrete object to measure an abstraction is a dualism which will solve itself without much ceremony. What is the ratio between an ounce of gold and a metaphysical idea? Describe the nature and intensity of a desire whose utility equals that of a Bank of England note. How many increments of happiness are treasured up in a kilogramme of silver? How many tons of copper are required to balance the amount of scarcity which exists in the world? We have been educated to the belief that gold will purchase all the objects of our desires for all time to come; therefore, with all our getting, we must get gold. I should like to know what we shall have to purchase gold with when all other commodities are gone. Under the unbroken rule of existing economic theories, the time must come when there will be gold enough to pave all the streets in Christendom, and when there will not be sufficient fertility in the soil, or vitality in our domestic plants or animals, to support a living soul. Many of us believe that if gold became as plentiful as stones everybody would be so rich that nobody would be obliged to work for a living; and those of us who have a little more intelligence are of the humble opinion that the multiplication of bank-notes, bonds, and mortgages is the true method of producing national wealth. Like other articles of wealth, gold must be scarce before it can possess economic value, and the scarcer it is the more valuable it becomes. Plentiful things never become fashionable, never provoke our desires, and the economic virtue of an article consists in its being scarce. Plentiful things may be made scarce by being subjected to a large amount of labour; hence the theory that labour is the true standard of value. One of my objects in this chapter is to show that neither money nor labour is the true standard of value, that the exact reverse is the case, namely, the more money and labour expended upon a commodity the less valuable it becomes; but as I have, in a previous chapter, confined the discussion to food commodities, I cannot go beyond this limit here. I shall not here deny that there are some commodities which are made more valuable by the expenditure of labour.

In contrast with the above-expressed conceptions of economic wealth, let us here introduce a few leading facts with reference to scientific wealth. The basis of our existence is food, or, rather, the land from which the food is obtained. Our very bodies and souls are also derived from some parcel or parcels of land. Land is therefore the basis of scientific wealth; and no scientist can ever dream of regarding money either as a basis of wealth or as a measure of its value. But the scientific and the economic idea of land is also very conflicting. The economist takes the acre as the basis of land, while to the scientist land is measured by the few handfuls of chemical compounds which the soil releases from year to year, producing a crop, and which gives the ground its entire value. Although it is hardly conceivable that the last dregs of soil-fertility can be drained by the removal of crops, yet every pinch removed and not restored tends in this direction, and is the primary source of wealth known to the scientist. What do gold and bonds and houses avail when the source of sub-sistence is gone? No scientist can conceive of a method of creating wealth by making his kilogrammes of chemical com-pounds scarce; in other words, they must be a positive quantity, not a negation or an abstraction. The sources of waste of scientific wealth may mainly be divided into three classes: (1) animal excretions (solid and liquid), (2) the bodies of animals, and (3) vegetable refuse. The last named also includes such sources as cast-off clothing, kitchen refuse, &c.; but heavy rains overflooding the land may also be a source of waste, although largely unavoidable, but this is compensated by floods which sometimes bring fertility from otherwise inaccessible quarters. In the state of nature no wealth is lost, for the excretions of wild animals are deposited on the land, as are also their bodies after death; so that the world is no poorer on account of their having existed. The posterity of wild animals has therefore no action for damages against their ancestors; but if the excretions were deposited in rivers or lakes, and their dead bodies in caves, the tran-sition would then be entitled to the name of "civilisation." Now, it is plain that if the excretions and bodies are deposited, nothing being wasted, the same number of animals could be produced and maintained for ever without drawing a particle of fertility out of the soil beyond that removed by the first generation. Therefore, if the sewers and graveyards of London were forced to deliver up their booty, another London could be produced and maintained, leaving the soil from which the inhabitants derive their subsistence unchanged in point of fertility. In this case, however, the cast-off clothing, kitchen waste, paper, framework of houses, &c., must also be restored to the soil, or, rather, the ashes of these waste materials. This is governed by the law of minimums, which will be explained in a later chapter. Now, what is remarkable about the situation is this, that our graves and sewers do not contain a farthing's worth of economic wealth, because there is nothing which appeals to our abstract desires; indeed, millions of economic wealth are being spent in dumping scientific wealth into watery caverns beyond all possibility of redemption. If we could calculate how many acres of land are being annually dumped into the sea, the figures would be alarming, even omitting the greater area wasted by the practice of maintaining domestic animals. It may here be argued that the hog, as the prince of scavengers, saves a great deal of rubbish from going to waste. All I have to say about this is, that the greatest of all scavengers is the animal which eats the hog, and yet this system of economy does not prevent waste. Should it be argued that the existence of cities makes such waste compulsory, then, as we shall see, science does not sanction their existence.

It is necessary in this inquiry to take a brief survey of money as the Zeus of our idolatry. Those who imagine that the basis of worship in any civilised country is in any other form of religion have a false conception of history. The deity of every age and nation is the dominant force by which they are controlled, the inferior forces being the minor deities. As we are now controlled by economic forces, we must seek in them the basis of our worship. Christianity (indeed all religions) has become a subordinate divinity, which cannot be completely eliminated so long as abstract theories hold sway. In the economic godhead there are three persons: (1) the money-made man, (2) the tailor-made man, and (3) the theory-made man. We worship him seven days in the week. Being an objective deity, he is not the God of the Jews, but corresponds to the God of those ancient heathens, the Greeks. He created us in his own image, and for his own glory and pleasure. He presides over manufacture and commerce, and gives us supernatural power to subdue nature and natural laws. In order to expand manufacture and commerce, he inspires his elect to Christianise the heathens abroad, and to prostitute our sisters and daughters at home. He is the measure of all things, human and divine. He is the God-king, and his anointed, vicegerents of heaven, reign and govern on earth by divine right—demigods, Czars of economicism, each having a civilised nation under his magic and majestic sceptre. He gave us dominion over all creeping things upon the earth, and made the herbs as meat for all creeping creatures. Excommunication, boycotting, and eternal damnation are the reward of all who make manifestations of uneasiness or show signs of disobedience under his Almighty sway, while a glorious future—a paradise—awaits all who worship him in spirit and in truth—through the encouragements of trade.

Those who "bemuddle the various schools of currency quacks"[1] are the only disciples of the orthodox school of political economy who are now making their voices heard. The new scheme to improve agriculture in particular, as well as business generally, is bimetallism, and the subject has already been discussed in the House of Commons, but with ill success. Gold sways the individual to-day as much as it swayed the nation during the rule of Mercantilism. Money must stand or fall with the rest of the economic fabric; it is subject to the same laws, and leads to similar divergences of thought and embitterments of feeling. In these days of the minute divisions of labour, it is somewhat surprising that money still continues to perform so many functions. Besides being the standard of the man, it is used as a measure of value through which different commodities may be compared; it is used for facilitating exchange caused by the division of labour, for estimating future contracts or obligations, and for hoarding up values. The qualities to be possessed by a commodity which is to serve as money are said to be as follows: (1) it must possess a distinct utility apart from its use as money; (2) it must be divisible; (3) homogeneous (like in kind); (4) cognisable; (5) portable. That gold and silver

---

[1] This high compliment to our currency philosophers has been recently paid by an editorial writer in the *Daily Telegraph* (Feb. 9, 1889).

possess these qualities in a higher degree than other commodities the scientist does not trouble himself to deny. As a medium of exchange, the function of money is to encourage trade, and this brings us back to the economic issue discussed in previous chapters. Why should we immolate ourselves on the altar of trade? Again, money is the measure of value. What is value? Economists have not been able to agree upon a definition of value or utility even in their own category. There is first value in use, and secondly exchange value, between which no effort seems to have been made to establish any ratio. Value is sometimes confused with the cost of production, and this cost is again conceived from two standpoints : (1) the cost measured in labour, and (2) the cost measured in money. Then we have absolute and intrinsic values, which are just as abstract as the other conceptions, and are therefore not susceptible of definition. In order to possess value, it is said that a commodity must subserve some human desire, and therefore must have utility or usefulness. What is the measure of human desire? Unless this question can be answered, and unless some ratio can be established between our desires and our destiny, all the theories of value must fall to the ground. Smith, Ricardo, and Marx base value upon the quantity of social labour expended, while Jevons, Wicksteed, and their school take the law of supply and demand as the basis, and this is supposed to be regulated by utility. Mill thought that labour was not always the entire source of value. The fallacy of labour being the standard of value is strikingly illustrated by the products of agriculture. Heavy crops do not require much more labour than light yields, and yet when the harvests are generally abundant the prices fall so low that an agricultural panic may result. The fact that an abundance of the necessities of life can pauperise any portion of the community should in itself be sufficient to upset all the theories of political economy, and proves the fallacy of scarcity as a factor in the production of wealth. It is sometimes urged as a condition of value that the commodities must be freely produced, the cost of production then being, in the long-run, the measure of value, but this condition can never be attained so long as all the land is monopolised. The theory that labour is the standard of value leads to the conclusion that the labourer gets the total produce of his labour, just as the rich man now gets all the produce and benefits which his money can command. This will give rise to an aristocracy of labour, which can never hold sway until all existing theories of education are revolutionised, and our rulers themselves must be labourers. Unless men can be found who are willing to labour and rule at the same time the theory is impracticable, for no man will want to disgrace himself by ruling so long as the aristocracy are all engaged in labour. It is specially inconsistent with Socialism, for under its rule officialism will be indefinitely increased. It also leads to the conclusion that labour, being an aristocratic employment, must exist in great abundance, so that human desire must be proportionately increased, and this is the basis upon which the whole fabric of economicism falls to the ground. No scientist can countenance such a scheme, his method of thinking being the direct reverse, and leads to the ultimate conclusion that labour must be abolished—that is, labour in the current meaning of the word. It is impossible to create an aristocracy of labour so long as respectability remains attached to the theories of ideas which form the basis of our educational institutions. Another weakness in the scheme is that the labourers cannot receive the whole produce of their labour so long as there are rulers, theorists, weaklings, unemployed people, &c., to be fed, clothed, and housed; and the same conditions which now exist must be perpetuated, one portion of the community wearing their lives away to support non-producers, no odds what the standard of value may be. So long as the sum of our desires remains unchanged, the quantity of labour required to appease them cannot be lessened. There must be some check to inordinate indulgence, and wealth must have some other basis than the defraudation of posterity for the purpose of satiating our licentious desires.

With reference to money as the standard for measuring future obligations, the question arises, Why should mankind be divided into two classes: (1) creditors (masters), and (2) debtors (slaves)? If gold is required for hoarding up values, then why should values be hoarded up? We must also have the poor man's farthing, the rich man's guinea, and the shilling for the middle classes. Why should there be rich men and poor? Oh, some people are diligent and others lazy; some are penurious and others prodigal. Very good. But do the diligent and penurious possess the most property, and are the lazy and prodigal always sunk in the lowest depths of poverty? Indeed, the ultimate analysis does not rest here: the scientist still wants to know why idle and improvident people should exist.

Taking food as the basis of our subsistence—and if we have too much raw material for its production, the surplus may be converted into clothing, or into trees with which we can build houses—let us now examine the pretensions of money and labour as a standard, measured by the scientific basis. In order to show more closely the scientific value and the economic price of those articles of diet which are of general consumption, I have prepared the following table of comparisons. For the principles involved in the table I am indebted to the indefatigable labours of J. Kœnig, one of the most distinguished of the German food chemists. He has compared the market prices of foods during a number of years, and found that those containing the highest percentage of protein brought the highest prices, and the foods rich in fat brought more than those rich in carbo-hydrates. The discussions on this subject, which have been specially fierce with reference to cattle foods, reacted on the market prices, and the relation of the nutrients now possesses high scientific value. At first the relative values were 5 units for the protein, 3 for the fats, and 1 for the carbo-hydrates, which relation has been pretty closely adhered to for articles of human consumption; but for stock foods, a gradual sinking in the relative values of the protein and fat has been taking place—or, rather, it has been found that the carbo-hydrates possess a relatively higher value, especially due to the fact that cellulose, which was formerly regarded as being very indigestible, is now known to possess a high feeding value, not only as having a high digestive coefficient, but also as having a mechanical value. In human dietetics the mechanical value of cellulose is also high, although the digestive coefficient is usually low, but the quantity consumed is usually so small that its consideration is a matter of little consequence. The relation in stock foods has recently been narrowed down to 3 : 2 : 1 respectively for the nutrients in the order already named. In cattle-feeding it is always understood that the food is consumed in the raw state, and I have no hesitation whatever in accepting the narrow relation for the scientific vegetarian, and the wide relation (5 : 3 : 1) for the economic vegetarian, for when cooked foods are consumed the relation must be wider, due to the coagulation and consequent lower digestive coefficient of the protein. It is argued that the digestibility of the starch is improved by cooking, due to the bursting of the granules, but this effect cannot compensate the advantages gained by the thorough mastication which raw foods must receive. Besides, the rejection of 3 units of value for the purpose of saving 1 unit is as irrational in principle as throwing away a penny in order to save a farthing. Having shown pretty clearly that biology is the basis of dietetics, not chemistry, and my inquiry being scientific, I am obliged to accept the narrow relation. If any objection be urged against this choice, the critic who chooses to figure on the wide relation will find that he can gain nothing. As my object is to prove the fallacy of money as a standard of value, not to preach dietetics, I have also included animal foods in the table.

Table showing the Scientific Values of Different Articles of Food Compared with Market Prices.

Names of Articles.	Chemical Analysis.			Units of Nutritive Vlaue.	Market Prices in Shillings per 100 lbs.	Ratio between Shillings and Units; also the Purchasing Power of a Shilling.
	Protein.	Fat.	Carbohydrates.[1]			
	Per cent.	Per cent.	Per cent.			
Wheat (crushed)	12.4	1.8	70.0	110.8	20.00	1 : 5.4
Wheat (whole) .	12.4	1.8	70.0	110.8[2]	8.33	1 : 13.3
Oatmeal . .	12.6	5.6	64.0	113.0	14.3	1 : 8.0
Beans . .	23.2	2.1	53.6	128.4	24.42	1 : 5.3
Rice . . .	8.0	1.0	76.5	102.5	16.66	1 : 6.2
Maize . .	10.0	5.5	63.0	104.0	7.15	1 : 14.5
Eggs . . .	12.5	12.0	...	61.5	64.00	1 : 1.0
Milk (whole) .	3.4	3.4	4.8	21.8	11.60	1 : 1.9
Milk (skim) .	4.0	0.7	4.9	18.3	5.80	1 : 3.2
Cheese (fat) .	25.0	29.0	2.4	135.4	66.60	1 : 2.0
Cheese (skim) .	35.0	11.5	5.5	133.5	22.20	1 : 6.0
Butter . .	0.6	83.6	0.4	169.4	116.66	1 : 1.4
Beef (fat) . .	11.0	23.0	...	79.0	83.00	1 : 1.0
Beef (lean) .	21.0	2.0	...	67.0	58.00	1 : 1.2
Bacon . .	5.0	78.0	...	171.0	75.00	1 : 2.3
Fruit (fresh) .	0.5	...	10.0	11.5	20.00	1 : 1.2
Fruit (dried)[3] .	2.5	1.0	55.0	64.5	38.00	1 : 1.7
Almond nuts (shelled)	24.2	53.7	7.2	187.2	92.00	1 : 2.0
Potatoes . .	2.0	...	20.0	26.0	6.25	1 : 4.2
Cabbage . .	2.1	0.5	6.5	13.5	8.33	1 : 1.5
Sugar . .	...	...	99.0	99.0	16.66	1 : 6.0

In the above table I have taken ordinary average retail prices in London for foods of fair average quality. This table has great significance in the solution of the social problem. Beef or eggs may be taken as the standard of comparison, for in both cases one shilling purchases one unit of

[1] The carbo-hydrates include sugar, starch, cellulose, and dextrin.

[2] This sum is obtained by multiplying the protein by 3, the fat by 2, and the carbo-hydrates by 1, then adding the products. By using the wide relation (5 : 3 : 1), the total units would be 137.4, and a shilling would purchase 16.5 units, instead of 13.3.

[3] Fruits have only a trace of fat, so that nuts are best adapted to make up the deficiency. The large percentage of sugar which fruits contain renders "jam" worse than useless. Nuts also supply the deficiency of protein in the fruits.

nutritive value, and they are the dearest of all the twenty-one articles of diet contained in the list, maize and whole wheat being the cheapest, and the other articles varying according to the figures given. The question now arises, Why do we pay a shilling per unit for meat or eggs when we can get 13.3 identical units of nutritive value in wheat for the same money? Putting it in another shape: If a shilling's worth of meat will support a person for one day, the same shilling will purchase sufficient wheat to last 13.3 days, and so on throughout the whole list, providing a selection be made which gives the proper albuminoid ratio; for example, nobody can live on butter or sugar, so that they must be eaten with foods containing high percentages of protein. The most striking feature of the table, however, is the fact that the dearest foods are, as a rule, those which require the most labour, are also the least adapted for human consumption, and require the greatest area of land for their production. For example, if we consumed the grain ourselves, instead of feeding it to stock, all the labour in feeding, tending, and housing the animals would be saved, and also, as we have seen, large areas of land would be saved. To discuss this question in detail the value of the by-products would have to be considered; but, on the face of the question, the margin is so wide that we can clearly see a ratio established between the three significant factors—land, labour, and money. The more labour expended the greater the quantity of land and money required, and the more useless and wasteful becomes the commodity. Therefore labour destroys values instead of creating them, and, as we shall also see, the labour applied to the land in the production of foods is also a destructive agency. In every way we turn we find that economic values are created by the destruction of scientific values, which proves the fallacy of abstract methods and the crimes to which they lead. If we examine more closely into the cause, we shall find that people consume expensive foods, not because such articles are wholesome or even palatable, but simply because they can afford to live luxuriously, must make a display of their respectability, and are ashamed to be seen sneaking into a vegetarian restaurant for a sixpenny dinner when they can afford to spend as many shillings in a grill-shop for the same number of units of nutritive value. Money and labour, as standards of value, must stand or fall together, and if they fall abstract education must tumble down as a component part of the same fabric. There is no possibility of separating material and moral problems: it would be just as rational to attempt to live after our bodies and souls are separated, or to perpetuate a vital organ when the structure and function are rent asunder.

Another significant lesson may be learned from the table, namely, that dried fruits are imported from Greece and other distant countries more cheaply than we can raise fresh fruits at home—that if we encouraged fruit-growing until the produce of our orchards and gardens became about 50 per cent. cheaper our fruit trade would be ruined, so that land monopoly, which prevents free access to the land, must be encouraged if trade is to flourish. It will not do to say that imports may be neglected and exports encouraged, for exports are settled by imports. The encouragement of agriculture and commerce side by side is the gravest form of lunacy. Civilised peoples are doubly heathenised: they destroy one another both in times of peace and in times of war, and their education consists in acquiring a technical knowledge of the instruments and engines of destruction.

In my observations concerning money I proceeded on the assumption that it was something which somebody could define. If money be the standard of value, it is necessary to know what money is; and even on this question our speculative thinkers cannot agree. Some confine it to gold; some include silver, especially under bimetallic rule; and other authorities would take offence if they were accused of not knowing that drafts, bank and other promissory notes, bills of exchange, &c., are money. A credit balance, one would think, should also be money; so should an honest and solvent man's word, which is said to be as good as the bank, and certainly serves the purposes of money. Some "currency quacks" think that gold cannot change its value, because it . could not be a standard if it were liable to change, but economic forces do not bend to this theory. Some regard gold (and

silver) as money because the coins are stamped with the image
and superscription of the king, whose fiat gives them their
value; while others are of the humble opinion that the metal
from which the coins are made must have the same value
before as after the Minting process. This latter view is very
disloyal, for it depreciates the omnipotence of kingly fiats.
Some insist that a currency system should be bimetallic, and
then arises a wrangle about the meaning of bimetallism.
Some think that trade would become brisk or brisker if the
State issued an unlimited quantity of paper money and set it
in circulation, while others maintain that the monetisation of
silver would be the best means of stimulating trade, for this
method would increase the quantity of money. Some who
have espoused the cause of agriculture contend that our
agricultural depression has been caused by the demonetisation
of silver, and urge as a remedy the return to bimetallism—all
civilised nations to unite and fix a given ratio between gold and
silver, and to make the latter a legal tender. Some are in favour
of utilising both gold and silver coins as standards, but making
the one independent of the other, no ratio being fixed; while
others advocate the use of coins composed of an admixture
of gold and silver in a certain ratio, and the opponents of
these two schools protest their adherence to the single gold
standard. Some insist on peremptory methods of reducing
our monetary system to a sound basis, while others think that
the change should be gradual, for when the ruled are robbed
by slow gradations there is less danger of an uprising, and
business confidence receives no violent shocks. These are a
few of the results of the attempts made to place our monetary
system on the soundest principles of political economy.

If our financial theorists could divest their minds of economic abstractions, they could not fail to see that gold (or silver)
possesses no virtue differing from any other article of commerce,
and that sovereign omniscience is just as potent in establishing
a ratio between the values of cats and dogs. If there is any
difference it is one of degree, not of kind. Many financiers
now admit this fact. Lord Liverpool says : " The Mint in-
dentures of Charles I., James II., Queen Anne, and even of
a part of the reign of George I., to the year 1717, had deter-

mined that the guinea should pass at the rate or value of 20s., and the other gold coins in proportion : yet they did not pass at that which was then their legal rate value, but at a much higher rate or value, and in a part of the reign of King William the guinea was current at even so high a value as 30s. This increased rate or value was not owing singly to a mistaken estimation at the Mint of the relative value of gold to silver, but the gold coins rose or fell as the silver coins were less or more perfect." During this period silver was the only legal standard in England, and gold coins were estimated by the silver basis, so that the guinea was 21s., a shilling not being the twenty-first part of a guinea, as now. Besides, the political power of the king, and even that of his advisers, was greater then than now. If the fiat theories had any force, their champions make a mistake in not advocating the monetisation of marble or granite, forcing the coins on the public at gold values. At least they should make silver as valuable as gold. The value of gold and silver, like that of all other commodities, is regulated by human desire ; and when ladies cease to improve their beauty by wearing gold trinkets, or cease to desire silver chains and collars for their poodles, these metals will become cheaper, other conditions remaining the same, and our bread and butterine will become relatively dearer, no odds whether the king be divine or human.

These constant tinkerings with our monetary standards have caused many and serious trade disturbances, and have been the source of much demoralisation and robbery. Who has not heard of bondholders' conspiracies to demonetise silver, or government conspiracies to enrich our rulers by debasing the coinage when they failed to effectually rob the people by various schemes of taxation ? Who has not heard of gamblers developing into millionaires through speculations in coins and bullions ? These public robbers, not mentioning those of Stock Exchange notoriety, are the people who are to survive and go down to posterity, for their victims, according to the neo-Malthusian theories of population, must not get married, because they are not able to support themselves and families. If economic forces are conducive to the breeding of thieves, why would not a reversing of the forces be conducive to the breeding of honest men? Are these desperadoes producers? Most certainly; they produce human desire, thereby enhancing values by making commodities scarce, which encourages trade, and furnishes work for the unemployed. They require enormous areas of land for their support, thus raising rents and increasing the wealth of the nation. Who has not heard of stock speculators listening with all ears to the king's speech or the president's message, in order that they may be able to calculate more closely the effects upon human desire? And when a panic is thus created, or business confidence shaken, a new religion is inaugurated by casting the blame on the sun-spots. Listen to the ejaculations of the bloodhounds of war when it is their interest to influence human desire and bull the stock markets. Listen to the wailings of the impoverished, as well as those who feel discouraged under business reverses or trade depression, when they subsist on hopes of commercial revival, and wait diligently for something to turn up—when money will again be abundant and freely circulated.

Many minor instances might be adduced to show the fallacy of money as the standard of value. Is the value of a commodity based on the wholesale or the retail price? Let us suppose three parcels of land, each containing a house; that the location is the same, or equally advantageous in every respect; and that the cost, measured in money or labour, is also identical in each of the three instances. The parcels, owned respectively by A. B. and C., are now for sale. A. advertises in the papers and realises, after clearing advertising expenses, £1000. B. not only advertises extensively, but also invokes the aid of the land agents, and nets £1200. C., who merely posts a notice "For Sale" on the front of his house, finds a customer for £800. Now, what is the real economic value or price of each of these parcels of land? The same rule applies to all marketable commodities. Here it must be admitted that business capacity, or incapacity, has something to do with the standard of value—the ability, or inability, to appeal to or awaken human desire. What becomes of the economic standard when a commodity is enhanced in value through the manufacturer's or trader's honesty or ability to supply the article pure and unadulterated ? What becomes of the economic standard when, to-day, the speculator purchases from the State a monopoly, and, to-morrow, the commodity is taxed up to the full value of the monopoly ? All our lands being monopolised—and the case is not altered by saying that they belong to the State—what is the true value of each product of the land ? There is no other conclusion than that the true standard of economic wealth lies in some ratio between monopoly and desire, and the economist who first establishes this ratio is the only man who can solve the social problem in the abstract category. Special values are subject to the same inconsistencies. For example, a picture or statue of a dying man bleeding on a cross might have great value, apart from the market price, for a blood-thirsty man in a bloodthirsty age ; but it would be utterly valueless in an age when human sympathies predominate. A noteworthy instance of economic value may be cited by the practice of substituting margarine for butter. The latter, although an utterly useless commodity, and costing much money and labour in its production, must have a place on our tables because the habit has become fashionable, we must "butter" our bread in restaurants even though the stuff be composed of the vilest filth, and we must often pay butter prices for stuff which costs a mere nominal amount of money and labour. Here we find that a sort of indefinable, tyran-nical, apistic abstraction called Fashionism is an important factor in the creation of economic values ; and this principle is still more aptly illustrated with reference to our habits of dress. It is that peculiar form of desire which impels us to cling to the "wisdom of our fathers."

Having shown the close relation between money and labour as a standard of value, it may be urged that it is illogical to say that commodities can be made scarce by the expenditure of labour. This question can only be decided by the basis which is assumed. Commodities are certainly increased by labour, but it must not be forgotten that the logical conclusion of this process ends in the destruction of commodities as form-ing a portion of wealth ; for when over-production reaches its ultimate the commodities have no value, and then cease to be commodities in the economic sense. There are two methods of increasing value : (1) restriction of the production, and (2) inflation of human desire. This is the real dualism in the problem relating to values. The untenable and illogical character of the position is this, that national wealth is created by the production of a great number of scarce commodities. The situation forces society into two classes; (1) the producers who make the commodities, and (2) the non-producers who, by inflating human desire, make them scarce. The former produce concrete, and the latter abstract, values, if values are produced. Here we have the concrete and the abstract, the material and the moral, phases of the question. It cannot be denied that the abstract phase is the basis of wealth, for political economy has been built up by men who have adopted the abstract method of thinking. The situation is aptly illustrated by the abstract conception of faith, or belief in miracles. Which is the basis of the miracle, the belief or the actual performance? Do miracles exist because a universal belief in them prevails, or because ample proof has been educed with reference to their performance? Those who have faith in miracles can easily understand how wealth can be produced by making it scarce.

Under economic rule, the man of money is monarch of all he surveys. He it is who encourages trade, causes money to circulate, and increases the national wealth. He is deified for the good he has done, even after his business collapse, and after bringing untold suffering upon many families, no difference how ill-gotten his pillage may have been. If I were asked who was the most selfish, omnipotent, despotic, bloodthirsty, and soul-destructive of all the Gods, my answer would be—GOLD.

But there is also another view of standards which belongs to this chapter, and which cannot be evaded either in economic or in scientific institutions. Everybody knows what is meant by measures of length (yard, foot, inch, &c.), and weight (pounds, ounces, &c.), and capacity (gallons, quarts, &c.), and value (£. s. d., &c.) ; but few of us have sufficient ability or learning to comprehend the inextricable muddle involved in our system of weights and measures, &c., especially when compared with the simplicity and efficiency of the metric or decimal system used on the Continent and by our own scientists, the comprehension of which demands neither learning nor ability. Many of us believe that our economic system of weights, measures, and currency is the simplest in the world, but this opinion cannot alter the fact, which may be tested in various ways. I mention these circumstances because my new system may be pronounced a failure if measured by English standards, while, under a decimal system, it might be regarded as a grand success; and there should be a relation between all measures, including time and force, as well as space, capacity, &c. Let two countries compete for commercial supremacy under identical conditions, except that the one adopts the decimal and the other the English system of reckoning, and the fate of the latter would soon be sealed. And yet we expect our commercial supremacy to be maintained! The decimal system of coinage adopted in America, Germany, and France would alone be sufficient to destroy our commercial supremacy under otherwise identical conditions; for under the former the business manager can make mental calculations to such an extent as to save the expenses of a clerk and the outlay of the materials used in making paper calculations; and even after this stroke of economy he has sufficient time left to enable him to speak politely and civilly to his customers, and to learn something about the quality of his goods. These facts prove that, even in the abstract category, science must seize the reins of economicism. It is a satire on our intelligence that the greatest economic nation should have the worst economic system. Indeed, men of the right stamp of intelligence are not those who delight in commercial speculations, or in developing the theories of political economy.

My position now is to establish a scientific basis of value, and to discover a form of wealth which no amount of scarcity can augment.

CHAPTER IX.

THE history of mankind is characterised as ebbs and flows of objectivity and subjectivity. These conditions, being the outcome of artificial education and circumstances, and bearing no relation to natural instincts, must continue until a scientific basis of human conduct be found and acted upon, or until our race becomes extinct. Of all animals, man is the farthest removed from the laws of his being and development, and those creatures which render the strictest obedience to these laws are the fittest to survive in the struggles for existence. The function cannot remain after the structure disappears, so that the human soul cannot be perpetuated after man becomes structurally extinct, and an atrophied structure carries a weakened function. Religion and philosophy are both abstractions, sometimes exhibiting an objective, and sometimes a subjective, phase. The Greek religion was essentially objective, and was conducive to the development of thought (philosophy, science, and art), while the Hebrew religion was essentially subjective and conservative, the outer world being a mere symbol of the inner consciousness. Descriptive poetry is characteristic of objective religion, while impassioned poetry is characteristic of the subjective form.

In my inquiry into the origin of religion, I have not introduced the scientific researches of recent inquirers who maintain that Christianity has no historical evidence to support it, for many of the most important questions may still be regarded as debatable. The strongest evidence against Christianity, the facts of which cannot be disputed, is contained in the history of the origin and development of all religions, and in the fact that the circumstances and necessities of the times irresistibly tended to restore unity and harmony of thought by some method of conciliating the dualism of objectivity and subjectivity; and, owing to the predominance of abstract methods of thinking, and to the widespread belief in miracles, natural laws were twisted into conformity with the prevailing ideas, so that Christ's Incarnation was intended as the realisation of the ideal, and his Ascension as the idealisation of the real, thus conciliating both the Jewish and the Greek methods. Recent research into the historical evidence of the writings contained in the New Testament tends to substantiate the proofs derived from other scientific sources. Christianity was just as much a necessity of the times as was feudalism in later centuries; but as the origin of the world can now be accounted for through the scientific method of thinking, all abstract speculations must, sooner or later, fall into disrepute and ultimate perdition. No deity can now originate who is not obliged to obey the laws of his being and development, although he may be vastly more highly developed than creatures upon our earth, the legitimate conclusion being that natural laws are immutable, and by obeying them we worship our creator, and evolve into deities—that is, we enter the Kingdom of Heaven. By this method we obey all deities, both him who is the highest, and those who are lower in the scale of evolution. The existence of a deity is conditioned by his obedience to the laws of his being and development.

It is desirable here to note the tendency of a few of the new religions which have been built upon the ruins of Christianity. Theism is a misnomer, its basis being philosophy rather than religion. Rejecting the Holy Scriptures as a truthless and an immoral book, it acknowledges the existence of an Omnipotent God who operates through natural laws, and his creatures attain a more or less perfect knowledge of him through the exercise of their reasoning faculties. What destroys it as a religion is, that it acknowledges no great prophet or founder—its absence of authority—and the Theist considers it to be every person's right and duty to think for himself in matters of religion. If there is any shade of difference between Theism and philosophy it is this, that what the former regards as a duty the latter regards as a

privilege. The Theists accept the truths of science to aid
them in attaining higher conceptions of the Deity, whom they
regard as operating through natural laws. If they carried
this truth to its legitimate conclusion, and thus obeyed these
laws, they would be Humanitists in the most exalted signification of the name. Their philosophy, which they call religion, can possess no element of perpetuity so long as they
operate within the abstract category, and they only go through
another cycle of thought which was, and always will be,
doomed to decay: indeed, when science is brought to the
aid of abstraction, the doom of the latter is all the more certain and impetuous. In Positivism, on the other hand, which,
in England, embraces the Religion of Humanity, we find the
religious basis; but it is not the religion of humanity—unless
the history of the past is taken as the basis of humanity. It
is the religion of Comte; it is Comteanity—just as Christianity is the religion of Christ. It is the old, old story of
ancestor-worship. A few examples will illustrate this truth.
The late William Frey, in an excellent little work [1] appealing
to the Positivists, says : " Auguste Comte, in his statements
concerning a flesh diet, merely expressed the opinion which
was unanimously accepted by all men of science at the time he
lived." Mr. Frey, as a vegetarian, renounced the flesh-eating
practice of the founder of his religion, and it remains to be
seen whether his fellow-Positivists will follow his example.
If so, the religious character of the Religion of Humanity
becomes weakened, and a precedent is established for the
renunciation of all the principles of Positivism. Let us now
take a glance into the future, say a thousand years hence, when
English will be a dead language, unless scientific reforms
in our institutions are grounded. The practice of meat-
eating will then, in all probability, be regarded as a relic of
barbarism. The philologists of those enlightened days will
tell us that meat did not necessarily include the flesh of
animals, that the expression " grace before meat" meant
before meals, whether there was meat on the table or not,
and the Positivist will then say that an all-wise being like

[1] Vegetarianism in Connection with the Religion of Humanity : Scientific Proofs.

Comte could not possibly have been so barbarous as to have eaten meat—that the vice was, therefore, not Comte's, but reposed in the false interpretation of history. Comte's divinity thus becomes established, and Comteanity then reaches the highest stage of subjectivity. Another question, however, arises which casts serious doubts on the divinity of Comte.

In the June (1889) issue of the *Nineteenth Century* there is an article entitled "An Appeal against Female Suffrage," which has the signatures of a large number of the lady aristocracy of England, and clearly bears the stamp of Positivism. Indeed, it is one of the tenets of the Religion of Humanity that women should not enjoy the franchise, and that the professions, as well as the trade of politics, are not fitted for women. But the time must come, in the interests of humanity, when the occupations will be made suitable for the people, and not the people for the occupations, as is the case under economic rule, and when no gentleman will follow an occupation which is a disgrace to a woman. What will become of Comteanity then ? There is only one conclusion, namely, that, like Christianity, it must sink into perdition. The lesser force must yield to the greater. Thus we find that, in the abstract category, neither religion nor philosophy possesses the elements of perpetuity. By devoting their energies to speculative thought, they drag us away from connection with the means of our subsistence, thus creating a class of slaves who must toil for the support of the aristocracy who are engaged in the framing of abstract fabrics—"castles in the air"—and the tendency of abstract methods to remove us from the laws of our being and development proves the suicidal character of the system. Those other systems, such as Agnosticism and Secularism, which are being built on the ruins of Christianity, are fundamentally destructive in their tendency. Their weakness consists mainly in the fact that their patriarchs have failed to recognise the truth that all religions must have their basis in the perversion of some natural instinct in man, so that all moral forces which are wholly destructive can never satisfy the longings and cravings of the human heart. In like manner, all philosophies are cartoons of our natural ideas. Our nervous structures have assumed unnatural functions, and no abstract processes can generate harmony. All abstract methods, be they religion, philosophy, economics, or politics, must go through the same process of decay, for they can only end in objectivity, and, by basing respectability upon the theories of ideas, can only result in class warfare—the creation of a democracy who produce concrete values, and an aristocracy who produce values in the abstract. When the real is based upon the ideal, the door is open for access to the means of subsistence by supernatural processes, as is aptly illustrated by the theories of population; but as nature only supplies food through natural channels, the constant pressure must end in the extinction of abstract methods, and in the development of science. In the final struggle, people as a body are impelled by starvation and the pangs of extreme bodily endurance to yield to the real, however strong their professions and protestations may be with reference to the supremacy of the ideal. An ideal basis of existence can only be maintained under conditions in which people will voluntarily sacrifice their lives to their ideas, but the instinct of race preservation is too strong to admit of the practicability of the ideal basis. In other words, scientific subjectivity is master of the situation.

The history of religion amongst civilised people proves the insufficiency of abnormal and unnatural forms of subjectivity. For instance, the Salvation Army of to-day is the Methodism of a century and a half ago. The latter was then a strongly subjective element, but since the denomination has become opulent and powerful, it can now afford to confine its duties within the bounds of economic respectability, and its material interests consist in proclaiming its property to be consecrated grounds, and thus inaccessible to mundane control. In other language, it is now essentially an objective institution—a property defence association—and its former duties have been usurped by the Salvation Army, which, under economic rule, must undergo a similar process of evolution. When a certain stage of objectivity is reached, the denomination becomes a political party, and strengthens the powers of the State. With every process of such evolutions the social situation becomes more deplorable, for the increase of all temporal and spiritual offices casts heavier burdens upon the real producers, and it is not therefore surprising that the suffering masses lose faith in our institutions and organise into a subjective element on the basis of humanity. Our old and venerable institutions which have been founded upon the theories of ideas have no charm for them, and science, which busies itself with stern realities, is their only succour. Even the dreadful theory of everlasting torment, which could only have originated in a bloodthirsty age, has failed to coerce us to submit to the sovereignty of the ideal. The scientific method must also revolutionise our ideas of faith ; and the time is at hand when it will be considered more respectable to attach our faith to the credible than to attempt to convince ourselves that we believe in the incredible. The theory that miracles must abound in order to quicken our faith has proved a failure in the stern realities of life ; and if it be urged that there is no virtue in believing the credible, just as there is no virtue in abstaining from theft when commodities are abundant, then we are confronted with the relation between words and the entities, or nonentities, which they are supposed to express. The existence of a word does not prove the existence of a corresponding entity ; neither does it follow that every non-entity has a word which can be utilised for its expression. If abstract entities, or the negation of concrete entities— such as scarcity—must abound in order that the word honesty may be retained in our language, then all I have to say here is, that this theory of ethics has proved a failure, and is opposed to scientific methods. If a new commodity be made scarce for the purpose of developing our wealth and honesty, it also follows that every newly discovered nonentity, accom-panied by a new word for its expression, and every fresh miracle essayed for the purpose of enabling us to quicken our faith, are progressive steps in the moral world. The philosopher who constructs a new economic system in the abstract category must arrange his commodities or his dis-ciples in such a manner that everybody is always on the point of stealing, and yet always refrains from theft, and from this basis he may be able to found a new school of morals.

N

Agriculture has been captured by economicism, and our agricultural authorities have committed the same blunder in accepting chemistry to be the basis of their science as our doctors have done in accepting it to be the basis of dietetics. As we shall see, the principles of nutrition and the law of development do not differ in plants and animals. The basis of agriculture and dietetics should be biology, not chemistry, and should there be an apparent conflict, the former science must rule, as Scholasticism had to submit to Theology when disagreement took place on technical differences. For example, the biologist will have no hesitation in saying that grass is the best food for cows; and if the chemist proves that bran or oilcake produces more milk, the biologist will say, "So much the worse for the cow." A cow may be stimulated by artificial foods and drugs to give an unnatural flow of milk, but such a process of feeding can only tend to the degeneracy of the cow and to the final extinction of the whole breed, especially when other animals compete in the struggle for existence. Economicism favours the chemical basis, for we are driven to draw on the future—that is, to rob posterity—in all departments of human activity, and so much so in our live-stock industry that the heating of cow-stables by means of stoves and the feeding of the most easily digested foods are now advocated, which deteriorate the quality of the milk and the indirect products of the cow—cheese, butter, &c. Exactly the same principle is applied to the feeding of plants, and here also economicism favours the chemical basis. In applying chemical fertilisers, the phosphates are made soluble, by means of which, under favourable conditions, the first crop gains an advantage in point of yield, but the vitality of the plant is impaired. The same objection applies to the use of nitrate of soda and sulphate of ammonia in place of the natural organic matter in the soil; also to the use of the soluble forms of potash, such as the sulphate and the chloride of potassium. Farmers are compelled to use these soluble forms of fertilisers, because, under economic rule, they must commit robbery, and draw on the future, in order to settle their high rents, or pay off their mortgages or other debts. In our economic system of dietetics, we also draw on the future both with reference to our own vitality and to the waste of soil-fertility, especially in our large cities which are the products of economicism.   Chemistry is an objective science and is under the control of economicism, while biology, the subjective basis, which has scarcely yet made its influence felt, is the only source of relief from all our economic ills.   The chemical basis of dietetics leads to the conclusion that mineral drugs can be utilised as nutritive and curative agents, which favours all the orthodox schools of medical quackery; while biology, which insists that man must live in harmony with vital laws, has a tendency to destroy the medical profession.   Unless the orthodox doctor can prove that his nostrums can make amends for violations of the laws of nature, his occupation is gone.   The theological theory that man is a supernatural animal, and is not, therefore, bound to obey natural laws, favours the bemuddled schools of medical quackery.   Medicine is not a science, but a developed form of alchemy—the development of an ancient superstition —just as our laws are the development of heathen philosophy and theology.   The biologist can make no effectual progress in his science until he removes the superstitions which cling to our systems of economics.   Before the scientist can make his influence practically felt he must conquer the fields of agriculture, economics, philosophy, and theology.   He stands alone against the extreme objectivity into which all civilised nations have fallen—an extreme which must, in the near future, bring about a reaction of some form of subjectivity. Our whole system of education is profoundly objective.   In the very essence of economic rule, it must be so.   Professor Tyndall, for example, estimates that nine-tenths of our educated classes are Balfourists, or adherents of abstract rights, while Professor Beesly thinks that nineteen-twentieths would be nearer the mark, and I dare say a very large percentage of them would, if they had the opportunity, out-Balfourise Balfourism in nineteenth century atrocities.   The Irish question is that of the rights of humanity rather than a Home Rule problem, and although Nationalism is inconsistent with Humanitism, we must not forget the force of circumstances.   Under a reign of justice and equality, the Irish would not be Nationalists so much as enthusiasts in the cause of humanity, and their supporters should not lose sight of the fact that the real struggle is for a universal principle. So far as the social problem is concerned, Home Rule in itself will have no greater effect than a change of government would have in France, or a return to Protection in England. The source of objectivity amongst our educated classes is not far to seek. In our higher and highest institutions of learning, the theories of "vested rights" in the privileged classes are still taught, and the legitimate conclusion is, that men of learning must also have a monopoly of the branches of education through which they derive their subsistence. The same principle applies to our professions, and the question is not what end these abstract rights or theories subserve, but how they can be made to influence human desire. Ancient institutions and ceremonies must therefore be conserved in order that men of education, with the therewith conjoined social standing, may enjoy occupations in keeping with their dignity and respectability; and the masses are taught to believe that such functionaries are useful as consumers of the products of their labour, for any slacking off in the consumption tends to throw the labourers out of employment, or, these occupations being abandoned, the educated would be obliged to compete with ordinary workers in the labour markets, thus reducing the wages. An increase of unbelief in religion, or of the moral or sanitary condition of the people, would, by the same method of thinking, have the same effects, for then clergymen, lawyers, and doctors would also be cast into the labour markets. The most heavily taxed commodities are those which tend to demoralise the people. All these evils and inconsistencies are the results of the theories of monopoly and vested rights, and of the utilisation of the ideal as the standard of our moral conduct. This abstraction is carried so far that many of the labouring classes are induced to believe that it is the duty of the unemployed rich to waste commodities which they cannot consume in order to furnish labour for the working classes. It is rather surprising that our economists do not advocate an increase of population to consume the surplus commodities in times of over-production; but, instead of doing so, they follow the abstract method of thinking by asserting that, when our markets are glutted, there is then over-population, because there are so many people out of employment. In their way of thinking, the number of the unemployed is a measure of the extent of over-population; and yet, if these surplus labourers were removed, our markets would be still more congested, for there would be less people to consume the over-produced commodities. Whatever way we turn, we find inconsistencies and absurdities, and such must continue so long as we operate from an abstract basis, or find ulterior dualisms to be solved. The condition has led us to despair, and to affirm, in the words of Bastiat, that " What is, is what ought to be " —in other words, " What is, is right"—the inference being that we should stop grinding economic theories. If " What is, is right," then what has been, has been right, and what will be, will be right: therefore, when scientific principles prevail, they will be right.

One of the best tests of a theory is the conclusions to which it inevitably leads. If, as some theorists maintain, a greater population can be sustained on animal than on vegetable foods, then the population may be still further increased there-through that the ox be first fed to the hog, before he is consumed by human beings, for the consumers, it is supposed, would then have both the ox and the hog. This theory might be carried out indefinitely: for example, the ox might be fed to the hog, the hog to the hen, the hen to the fox, the human beings then consuming the fox. When all the beasts of the forest and the air thus became domesticated, and if the population still pressed against the means of subsistence, a remedy for over-population might then be introduced by feeding the final animal under domestication to the worms, the latter being used for human consumption. Nobody having ever proposed to try this experiment for the increase of population, or for the relief of over-population, the theory should fall to the ground. Equally absurd is the theory that the fertility of the soil can be increased by the manure from domesticated animals; and yet this theory is accepted by all agricultural authorities and practical farmers. The soil's fertility could not be increased by this method, even if all the manure and all the carcasses of the animals were restored to the soil, whereas it is well known that in practice from 50° to 75 per cent. of the manure and 80 or 90 per cent. of the carcasses are wasted. Granted, for the sake of argument, that the fertility of the soil can be increased in this manner, then what is the legitimate conclusion? It is this, that if human excrements were saved by being utilised, say, on sewage farms, then instead of the proprietors of hotels and boarding-houses receiving board-money from their guests, the balance due would be in favour of the latter, for they returned more values than they received. However absurd this conclusion may appear, yet it is identical with the theory that domestic animals must be kept merely as manure machines. But these conclusions are only introductory to the main absurdity. The motto of our Royal Agricultural Society is, "Practice with Science." This method of thinking could only have originated in an age when abstract ideas predominated. It has, for example, been ascertained by practical farmers that the *productivity* of the soil can be temporarily increased by manures and soluble fertilisers, and science must therefore be made subservient to practice; but there is a great difference between fertility and productivity. In like manner, it has been ascertained through practice that there is more money earned in ruining the health and vitality of domesticated animals, for the purpose of obtaining fictitious prices, than in the adoption of natural methods of improvement, and here again science must be made subservient to this cruel and suicidal practice. Our agricultural authorities fail to recognise the fact that true practice is the basis of their science, and in order to bring out this truth more clearly, I will show that the Royal Agricultural Society of England are not carrying out their own motto except so far as it applies to economic practices. Before the era of domestication, all plants and animals must have struggled for their existence just as wild animals and weeds have to struggle now. The practice of many thousands of years has taught us that vitality under domestication has become impaired—and this scientific truth applies to man himself as well as to domesticated plants and animals—and yet all the economic practices, which scientific agriculture is made to subserve, are a direct violation of the practical experiences taught by history. We thus find that there is a sharp line between those practices which form the basis of science and those which form the basis of economics. In the sense that science is drawn from practice, the motto "Practice with Science" is a grand truth; but it is utterly false in the method and interpretation of the Royal Agricultural Society. Using science in a wider sense so as to include true practice, from which it cannot be separated, the motto should be "Science with Practice"—or, rather, science with art—and when the Society recognise and act upon this truth our agriculture will flourish, but the attainment of this end is an impossibility so long as the abstract theory prevails that commerce is the basis of national prosperity. England, who promoted the economic idea, would, other conditions being identical, be the first nation to suffer the consequences of her folly; but her insular position, which is a wall of protection, has a modifying influence. Already our agricultural districts are rapidly becoming depopulated, our people are flooding the cities in search of employment, the London County Council is encouraging this tendency by a system of cheap and comfortable dwellings for the poor; and a great city may be defined as a target for the fiends of war. We boast of our expanding commerce, which is the same thing as to boast of the destruction of our agriculture, and it is our system of land monopoly which has made us "Mistress of the seas." If commerce were the true basis of national prosperity, this condition would be desirable; otherwise national disaster and collapse must be inevitable. Through our trade ideas we lost the American colonies a century ago, and now the same ideas are alienating Canada from the mother-country and driving her into the American Union. Our colonial possessions have virtually ceased to be of advantage to us as an outlet for our manufactured goods, for they have assumed the right to regulate their own tariffs and protect their own industries. They are ruled by the economic idea, and every generation witnesses a decline in that "United Empire Loyalty" which inspired the bosoms of their fathers. While all this is transpiring on the one hand, we go on conquering

other trade centres amongst the barbarians under the pretext
of civilising and Christianising them, but our resources cannot
stand this wasteful process much longer. There is a movement favouring the encouragement of our agriculture by State
aid, which must prove a lamentable failure, especially so long
as it remains a conspiracy for the enhancement of agricultural
rents. The conversion of the soil of the United Kingdom
into a garden under the system of peasant proprietorship
would be but the beginning of the evils of landlordism.

One of the greatest fallacies of our century is the theory
that war can be prevented by moral instrumentalities. The
Socialists in particular are deeply impressed by this idea
of universal fraternity and peace. Under economic rule, it
would be just as rational to attempt a reversal of the course
of the Niagara Falls. So long as commerce remains the basis
of national prosperity, or economic practice the basis of
science, so long as scientific wealth continues to be wasted
and economic wealth increased, and especially so long as
those who wage the wars and those who do the fighting belong to different classes, there will continue to be starving
millions and aggressive nations, and the State will be invoked
to protect itself against their encroachments. Any new
system which inspires public confidence may give temporary
relief, as a period of " commercial prosperity " does now, but
it is during such periods that the evils are hatched. Such
forms of prosperity are mere ideas which have no basis in
realities. Half-a-century ago Free-trade was the coming
millennium ; we are now in the midst of it and still we
do not seem to enjoy it. The coming millennium now is
Socialism, but as the controlling forces are not changed, the
same results must be anticipated. A century ago, the Rights
of Man were to be the millennium of the future, and we are
not even satisfied with that, although we have enjoyed it
for so many decades. Individualists, who are an organised
property defence association, cannot voluntarily give up their
property for the relief of the distress amongst the masses
without adopting the Socialistic idea, and if they wait until
they are obliged to relinquish their hold, the machinery
required for such a purpose can only be a tax upon them-

selves, and by virtue of their resistance they cannot hope to be classed in the universal fraternity which is to be established. The universal sentiment of humanity is destructive of the Individualistic idea in any of its modified forms. The system has committed suicide. Under Socialism, it sounds very well in theory that every individual or family should draw his revenues from the State coffers, but it must not be forgotten that the funds have to pass through the hands of economic and designing men, that there is no likelihood of creating employment for all, much less securing remunerative wages, and that it is impossible for each to obtain the full produce of his labour, or its equivalent. Even should the State undertake to inflate human desire to such an extent that ample employment would be guaranteed to all, the machinery required for carrying out such a scheme would in itself be sufficient to rob the people of much of the produce of their labour. An aristocracy of labour cannot be established until all our existing ideas of things are completely revolutionised; for it implies that people will be compelled to hold official positions or engage in literary employments just as they are now compelled to engage in manual labour; and implies, moreover, that the amount of manual labour to be performed is as inexhaustible as literary and scientific research. Mental labour can be performed without robbing posterity, but manual labour never. The Socialistic millennium, so long as the operations are continued within the abstract category, must suffer the same fate as all previous millenniums. The direct source of the social trouble may be summed up in a word : the systematic robbery of posterity. We must not forget that we are the posterity of our ancestors. This robbery may be classed under two heads : (1) the raw materials required for their subsistence, and (2) the vitality in the organic products of these materials. These are the results of economic forces. Omitting the soil's fertility, of which I have already spoken, our coal-beds are now mere matters of two or three generations, and our timber-supply is rapidly disappearing, but so long as the remainders bring an ever-increasing sum of money, we are induced to believe that our national wealth, in coal and timber, is on the increase. In like manner, wealth is created by wars, which strengthen the powers of the State, and posterity is obliged to shoulder the burdens. The force by which posterity is coerced to bear these burdens is ancestor-worship—the monopoly of our minds and consciences. In plain words, we criminally produce offspring, heap debts upon them, rob them of the means by which they are to liquidate these debts, deprive them of sufficient vitality to appropriate the means of subsistence for their own maintenance, and then wage war against them to reduce their surplus numbers and coerce the survivors into a condition of contentment; and the victims must undergo a strict medical inspection in order to determine if they are fit subjects to be killed. This is the economic law of the survival of the fittest. As a compensation for these burdens, posterity receives the blessings of civilisation. We (or any other generation) have only one recourse : we must either defend the cause of the dead or that of the prospective living. Any generation which is so selfish as merely to defend its own cause can only rob the prospective living, without doing any honour to the dead, and any attempted solution of the social problem on this basis will unceremoniously precipitate disaster. Indeed, the social aggravation may be expressed in this way, that each living generation has espoused the cause of the dead against posterity. The basis of humanity has been the past instead of the future. And yet, as I have shown, we have made posterity the pliant instrument for giving economic values to the commodities of which it is deprived. We have been driven into these economic extremes of objectivity with as much violence as our ancestors were driven into Feudalism and Christianity. If we do not put our shoulders to the robbery wheel, our neighbours will have all the better opportunity, so that we are obliged to plunder in order to prevent our neighbours from doing so, and they are obliged to do the same thing for the purpose of checking us. This is the whole law and gospel of social ethics and economics. The various new religions and philosophies which are being built upon the ruins of Christianity only tend to a greater divergency of human thought, and the same tendency appears in matters pertaining to our material well-being.   The time may not yet be ripe : it may require two or three more periods of "commercial prosperity" to convince us that economicism must collapse by virtue of its own innate rottenness, and to drive ancestor-worship to its objective extremity ; but it is to be hoped that there still remains in the human breast sufficient conscience to awaken our perishing souls to a sense of their accountability, and in our minds sufficient strength to extricate ourselves from the deep and dark abyss of objectivity into which we have fallen.

Is man degenerating?   This is a pertinent question in view of the subjects we have been discussing.   Economically speaking, we are· progressing, which fact cannot be denied, and so long as we are making progress, it is difficult to understand how we can degenerate.   Here again we strike a dualism between the abstract and the scientific method of thinking. Scientific progress cannot be disjoined from the forces which lead to the perpetuity of mankind, whereas economic progress is bound up with man's extinction.   All depends upon the category in which our thoughts move.   If we are on the road to perdition, we are now making satisfactory progress ; and those who speak of moving in other directions, or in solving the social problem, should, one would think, be peremptorily suppressed.   On the other hand, if our destination is at the terminus of the road which leads to salvation, this is another question, and there are grave problems to be solved.   But it is said that we are making scientific progress—that this is the age of science.   But, as I have shown, science has been helping us on our way to perdition, and if this be our destiny science is doing commendable work.   This fact I have shown in many ways, notably in making vital laws subordinate to those of the physical world, and in the tendency of new religions to use science as a stepping-stone in the attainment of higher abstract ideas.   The solution of the social problem may be said to consist in science rending asunder the fetters of economicism, as it has done with religion.   What we want is subjective science, not objective—that is, we must search in order that we may act, and not for the mere purpose of acquiring knowledge and of winning the applause of the economic world.   Scientists made a progressive step when they freed themselves from the fetters of the Church, but they are still bound firmly in the chains of other forms of abstraction. It may be urged that the adoption of scientific methods is a proof of mental progressiveness, but this argument bears a fallacy. People's minds have become scientific through the failure of abstract methods to produce unity and harmony of thought, and to create contentment in matters pertaining to our material wants, rather than through a natural predisposition in this direction. It is the outcome of circumstances, and any changes of this nature may take place under authenticated proofs of mental degeneracy. Under commercial prosperity, for example, economic theories tend to mould into a science and to produce harmony of thought, whereas under lengthy periods of commercial stagnation and depression there arises a tendency in the opposite direction. The fact that the philosophies and religions of to-day are not superior to those of ancient civilisations, despite the experience of two or three thousand years, is an unfavourable commentary on mental progress in the region of abstract thought. It is not just to say that our knowledge of economics is superior to that of the ancient Greeks, for they condemned economicism. The progress of invention is also the outcome of necessity, or of the circumstances of the times, and when a new idea once originates, its development is a matter of course. All these changes may take place under mental degeneracy. If we enjoy greater comforts than our forefathers, then these are mere abstractions, as I have illustrated by reference to temperatures, and tend to physical and mental degeneracy. We have even great politicians in our day, and it need not be denied that they have developed a certain amount of political and economic cunning, but this is no proof of an increase of general intelligence. To come to the point at once, it is a dualism to say that mental progress can take place under conditions of physical degeneracy; it leads to the illogical inference that an atrophied and weakened structure may be conducive to strength and activity of its function. If this be not the case, we must follow the abstract idea that mind may have an existence independent of nervous structure, which leads to the theologic theory of unfunctional immortality. It is said that the human brain is increasing in size, relative to the size of the body; but this, even recognising it as a fact, proves nothing, for it would be equally true to say that the body is decreasing, relative to the size of the brain. This change of relation may have a temporary existence; but, viewed from the standpoint of the logical conclusion, it cannot possess the element of perpetuity. A great deal has been said about longevity as an element in human progress, and I do not dispute its claims as an important factor. The few statistics, however, which have been gathered showing the octogenarians or centenarians in different ages of the world prove little or nothing. Our allotted span of "threescore and ten" seems to apply to the average duration of life rather than extremes of longevity. The fact that Abraham died at the advanced age of 175 years is unassailable evidence of our degeneracy, and the lives of other antediluvian patriarchs substantiate this position. Zeno, the founder of the Stoics, whose diet was simple, consisting mainly of bread, figs, and honey, is said to have died at the age of ninety-eight years, and even then did not die a natural death, his days having been ended through violence. This is strong presumptive evidence in favour of a simple and vegetarian dietary; and history records that Pythagoras (about 500 B.C.), who was the father of vegetarianism, having observed that man had nothing in common with carnivorous animals, so far as dietetic structures and functions are concerned, married at the age of sixty years, and afterwards raised a family who were the pride of their country. Viewed from another standpoint, the evidence of our degeneracy, so far as longevity is concerned, is complete. Amongst the lower animals, the natural duration of life is five to eight times the period between birth and maturity, by which standard man's age should be at least a hundred years; but the average has been reduced to less than twoscore. It is quite probable that prehistoric man lived several hundred years. Indeed, historical facts are no longer required to prove our degeneracy, physical and mental, for it is now well known that such is the inevitable tendency of all organic life which is removed from the law of its being and development, and as man is farthest removed from this law, he must become the unfittest to survive. Granted that each century produces its man of genius and courage—a man who has rendered tolerable obedience to the laws of nature and has profited by the experiences of past generations—yet this cannot alter the destiny fixed by general disobedience. With physical and mental degeneracy, moral and spiritual ruin must follow in the train. Show me the man whose ear is deaf to the tales of human woe, and whose heart is callous to the piteous sights of suffering and distress—the man who delights to vindicate the theories of abstract rights—and I will show you a measure of moral decrepitude.

The evolution of economic happiness is plain and simple. The happiness of the economic man can have no other basis than that of money-making—that is to say, the power of drawing on large areas of land for his support, and of enslaving those who labour upon the land or in connection with its commodities. Money is a land title. If money-making is the basis of happiness, money-losing must be the basis of economic misery. Admitting that it is revolting to the nature of the true scientist or the true artist to be under the painful necessity of making money, we are still confronted by the theory that the greatest happiness must accrue to the greatest number. It pricks the humane heart to the quick when the possessor thinks of or sees the butcher's dagger being thrust to the hilt into an innocent lamb bleating for mercy; but he must suppress his feelings, knowing that the operation affords such intense joy and pleasure to those economic gourmands whose teeth are watering in expectancy of the delicious repast. It pains him to see the wild hunters, born and educated in a civilised country, chase the bleeding stag behind the thicket where it oozes its last gush of innocent blood. It pains him to see the " birds of sweet melody " and the birds which prey on the enemies of vegetation ruthlessly slaughtered that their plumage may adorn the head-gear of the wives and daughters of our udæmonian philosophers. Are all these passions to be despised in order that the greatest happiness may flow to the darling souls of the greatest number? Must misery and wretchedness abound that we may glut our senses with the joys of our opulence? Yes; and it is the imperative duty of the State to see that the greatest number enjoy the keenest intensity of happiness.

We have now seen, briefly though pointedly, the failure of abstract methods in the solution of the social problem, and the necessity for a new basis of inquiry. We have seen the growing divergences in human thoughts and sympathies; and, in the attainment of unity and harmony, there is only one alternative, namely, the scientific method. There can be no compromise.

Having now completed the analytical chapters, or PART I., of this work, it is now in order to proceed synthetically in the following pages and develop scientific institutions in sympathy with the basis already laid down. We have seen that, in their ultimate analysis, all social problems are a land question, and that scientific titles must be based on the area of land which the individual draws on ~~for~~ for his support. This basis may be quite theoretic—that is, we may conceive the parcel of land as being located in one spot, and measured by the quantity of its products, but this does not interfere with the right of the individual to exchange the products of his parcel for those of other parcels, or with his right to permit other individuals to work his plot under conditions which are mutually satisfactory. In this manner the individual who does not work his plot must always receive the products or their equivalent, and can therefore never be reduced to a condition of want. Exchanging the products of one plot for those of another does not interfere with the practicability of reducing the products to their equivalents in land areas, and the area of land which a person draws on for his maintenance can be calculated with equal facility. For example, land in the United Kingdom yields 37 bushels of oats per acre, but in more favourable countries for the growth of this cereal the yield is about 60 bushels per acre, so that by an exchange of commodities the person who consumes large quantities of oatmeal will gain an advantage if he can make suitable trade arrangements. This is equivalent to the exchange of a portion of one plot of land for that of another, and trade will be forced into natural channels. There can be trade without economicism, and without any attempts to make it the basis of national prosperity. The question to be developed in the following pages is how to parcel off these plots in such a manner that the individuals will have a sufficiency to suit their varying requirements. Each individual now having a sufficiency of the good things of this world for the maintenance of himself and his family, the question may arise how and where he is to get the products to be hoarded up for his "heirs and assigns for ever." The answer is simple. If the country is fully populated, there can be no other plots which can be utilised for such a purpose; and, above all, there can be no inducement for him, in case of under-population, to work on plots which belong to succeeding generations, for, all monopoly being destroyed, each individual in every generation has equal opportunities for access to his plot of land. Besides, this is the only method for attaining an equal distribution of property, and for giving each individual equal opportunities for exercising his functions. The quantity of labour must therefore be regulated by that required to work a plot of land, or its equivalent in other occupations; but it does not follow from this that there is no other work to be performed: what I mean is, that the plot forms the basis of labour, and extra labour can be performed without wasting plots. With reference to the products of the mines, it is only necessary to establish some ratio between them and the plots of land. This, then, forms the foundation-stone of scientific economics, and all institutions must be in sympathy therewith.

In perusing the following part of this work, two precautions are necessary. (1.) Should it be found that the Scientific State has an Individualistic or a Socialistic bias, which forms I have condemned, it must not be forgotten that I am operating in quite a different category. It makes a world of difference whether agriculture, manufacture, and trade are made subservient to the welfare of mankind, or whether our race is to be sacrificed on the altar of these industries. These are two different categories—the concrete and the abstract. (2.) Any proposed changes must be evolutionary, not revolutionary, for no scientific methods can be revolutionary. A scientific system must grow; it cannot be made : the time required varies with the intensity of the change and the nature of the circumstances, just as the momentum is the product of the velocity by the mass. I make this observation here in order to evade unnecessary repetitions.

PART II.

THE REMEDY.

CHAPTER I.

THE EVOLUTION OF MAN AND NATURE.

GRANTED that science can present no direct proof of the origin of our planet, or of the material universe, yet a knowledge of the development of phenomena is sufficient for our purposes, and the hypotheses which are logically connected with the results of scientific investigation tend to the attainment of mental harmony. In the preceding pages, I have acknowledged the existence of matter and force, which dualism will be discussed in a later chapter, and we may take our starting-point from the hypothesis that our planet was originally a nebular mist, although any disproof thereof will not affect our inquiry. It is only necessary to go back to the pre-organic condition of our world, whence I hope to be able to evolve scientific wealth and destiny—the material and the moral.

Apart from the scientific method of thinking, as we cannot conceive of space being walled in, we must regard it as being illimitable, and not being able to conceive or find an absolute vacuum, space must be charged with matter, and force being inseparable therefrom, matter and force[1] are illimitable, indestructible, and eternal.

Force manifests itself in the forms of heat, light, electricity, magnetism, chemical affinity, and motion, any one thereof being convertible into any other, but no destruction can occur. All forces may be derived from motion, and may be actual or potential.

---

[1] It would seem to be just as correct to say that space is charged with force, matter being inseparable therefrom; but science takes matter as the basis of force, the force basis being the abstract method, which has led to all the fallacies which I have exposed. My method is to move from the abstract to the concrete category, making the structure the basis of the function.

Our primordial mist having been consolidated by slow gradations, let us carry our minds back into the azoic age of geology, before the appearance of vegetable or animal life upon our planet, and we behold matter in the forms of water and igneous rocks, with the various mechanical forces which act in and upon them. We behold huge glaciers, charged with rocky fragments, moving down to the lower levels and grinding the mountains into powder. We behold the streams pulverising their banks and beds, carrying the resulting *débris* to the bottom of the ocean, or depositing it upon inundated plains. We behold the rain descend, forming streams and streamlets, and we behold the self-same water again ascend in vaporous condition, forming clouds, and again descending in the shapes of snow, ice, or rain, the glaciers again being fed—all due to the heat of the sun. Even the soil, produced in this manner, is again converted into rock under the influence of heat, pressure, and other forces, and thus we find all nature repeating itself—from clouds to ocean, and ocean to clouds ; from rock to soil, and soil to rock—all in endless succession.

Can we here gain any conception of wealth or destiny? Certainly we can, however crude it may be. It is the destiny of the sun, so far as it affects our planet, to produce these changes ; it is the destiny of the glaciers to make soil, and it is the destiny of the soil to be turned into rock ; it is the destiny of the water to form clouds, and so on—each cycle of change appearing to aggregate no change—and neither matter nor force is produced or annihilated. All these changes are wealth produced by the sun ; the commodity of wealth which the glacier produces is called soil, that which pressure produces is called rock, &c. If force demands wages for exerting itself upon matter, then matter must have an equal amount of interest for lending itself to force.

Passing now over myriads of ages into the geological era of organic life—and for the logic of our inquiry it makes no difference whether plants and animals had a simultaneous origin or not—we enter into a more complex stage of wealth and destiny. The soil now produces the plant ; the plant dies and moulders into soil, and the soil returns to rock. It is the moral destiny of the plant to live and breathe and die, first reproducing its kind. The vegetable kingdom itself is the product of those ever-changing forces which gave origin to the various commodities in the pre-organic world: existence and fatality are alike the lot of organic and inorganic nature. It is now the manifest destiny of inorganic matter and forces to subserve inorganic life. Before the existence of the plant, it was the destiny of the inorganic world to bring about such changes as gave birth to organic forms, just as it was the destiny of vegetable nature to bring about conditions favourable to animal life. The vegetable is now the wealth produced directly by the mineral, and the latter has no more claims for compensation than the former has for the loan of its vitality.

All matter and force which tend to produce plant life are wealth-producing agencies. Of the various species of plants, each must be adapted to the environments under which it exists, and the original form of life must have been in harmony with the conditions which gave it being, which circumstances are essential to its existence. In other language, organic life must obey the law of its being and development. This is the scientific method, whereas in the abstract conception it is the function of organic life to change this law— that is, science is subjective, while abstraction is objective. Chemistry favours objectivity, and biology favours subjectivity. By the subjective basis, the changes which organic life undergoes follow universal laws, whereas by the objective basis these laws are subservient to the organic changes. Subjectively, organic life is enabled to exercise objective sovereignty over all forms of life whose obedience to the laws of their development is less pronounced: objectively, this order is reversed. This law of obedience not only applies to the plant as a whole, but also to each tissue of its structure, so that one tissue cannot perform the functions of another, except through such slow gradations as may be brought about in sympathy with and obedience to the law of development.

Let us now pass on to the animal kingdom—the ages of the molluscs, fishes, mammals—meanwhile omitting man from our inquiry. It is impossible to draw a sharp line of demarcation between plants and animals, there being so many characteristics in common, but, regarding the most distinctive feature to be *sense*, we now find matter and force occupied in the production of sense commodities, for which previous conditions were destined to prepare the way, and no change appears in the law of development. The sense function is inseparable from the structure, the latter now, in the scientific method, to be taken as the basis of measurement. The seat of sense is nerve tissue, before or after the formation of this structure, but this tissue, again, requires other structures for its base. The law of development of an organ or structure, which has already been pointed out, consists in the exercise of its natural function, and the morality is inseparable from the fulfilment of destiny. This law applies to all organic life. The destiny of the animal is to live and breathe and perish; and its bones are to produce other animals endowed with greater powers of sense. The plant, finding its food constantly manufactured beneath its feet by chemical agencies, requires neither intelligence nor storing respositories, while the animal, having to seek its food, requires both as a condition of its existence; and the intelligence, which is referred to the same structure as sensation, varies with the methods of procuring the means of subsistence. The fox, for example, has not developed his cunning because he is a carnivorous animal, but because his existence depends upon his success in developing this faculty—or, rather, the nature of his method in procuring the means of subsistence forces him to exercise his characteristic faculty. If the conditions demanded greater cunning for the procuring of grass than the capturing of the fowl, the sheep would be more cunning than the fox; otherwise it would cease to exist.

In order to understand what is meant by the law of evolution, so ably expounded by Darwin and now supported by all biologists of repute, let us here take a glimpse at its classification, namely :—1. Physical environments. 2. Increased use or disuse of organs due to change of environment. 3. Natural selection. 4. Sexual selection.

When an animal changes its surroundings, as living under different temperatures, procuring different foods by different methods, &c., changes of structures and functions are the natural sequences, which changes are proportionate to the variations in the environments and the period of time elapsed; and the modifications, together with the innate predisposition to exercise the most highly developed structures, are ultimately inherited by the offspring. Sufficient time and change of environment being given, there can be no limit to the modifications of structures and corresponding functions, so that an animal may easily evolve into what the old school of biologists call a different family or species, and if it came again into contact with the original stock, which either underwent no change of environments, or developed under essentially different environments, cross-breeding would become impossible. Geology has proved the great antiquity of our planet, so that these two sciences, geology and biology, harmonise so far as the transmutation of the species is concerned. Fossil remains prove that forms once existed which are now extinct; and the disappearance of intermediate species causes still greater variations in plants and animals. Under these changes, and through the influence of natural and sexual selection which also cause modifications, the fittest survives in the struggle for existence; and the tendency of all organic forms to multiply beyond the means of subsistence intensifies this struggle. This dualism of pleasure and pain, namely, the over-production of offspring in relation to the means of subsistence, is the basis of development and perpetuity in all organic life. This law of nature is important as representing the difference between the scientific and the abstract method. In the latter, it is assumed, without proof, that our planet and every species therein contained is a special creation : it starts with the idea and never gets beyond that limit, while science reverses this order, starting with the concrete and climbing up. Abstraction commences with the Deity ; science ends with Him. It is not, therefore, surprising to find a clashing between the two methods in all the problems of life. All attempts to complete a structure by commencing with the roof must prove a failure.

It is now, I hope, made plain what is meant by natural law, so far as it pertains specially to this inquiry. All organic life must do one of two things : (1) obey the law of its being and development, or (2) give place to forms which do. The transmutation of plant into animal life may be essentially characterised as the evolution of sensibility, and of simple into more complex structures and functions. All organs are developed by exercise in such movements as enable the animal to obtain its subsistence, and all modifications take place by tardy gradations and in conformity with fixed laws. The conditions of existence are twofold : (1) the subjective impulse through which the offspring is begotten and preserved, developing into an incentive for the preservation of the whole species; and (2) the objective impulse to perform such actions — and thus developing such structures and functions—as are requisite for the maintenance of the animal and its young. Were any of these conditions wanting, there would be no animal life; and the keener the struggle for existence, the greater becomes the necessity to render obedience to these laws—or, the more these laws are obeyed, the keener becomes the struggle for existence. Subjectivity thus forms the basis of objectivity; for the impulse to beget and preserve takes precedence, just as the sensation of hunger must exist before exertions are made to appease the appetite.

The feature of our planet up to this stage is that there are no dualisms—all is in harmony—and there is only one phenomenon, namely, nature. The dualism of pleasure and pain, which I have pointed out, and which I shall hereafter discuss, is not a dualism in any sense in which the word is now understood. Passing now on to the era of man, we are beset with the first and greatest of all dualisms—man and nature. Before this great epoch there was neither good nor evil in the world, and no supernatural agencies as distinct from the natural. For the sake of clearness it will be necessary for me to call everything natural which was created up to the evening of the sixth day, or before the creation of man, for man appears to be the only creature endowed with supernatural gifts, which, presumably, is the force which gives him dominion over the beasts, and enables him to subdue them. The difficulty in the way of understanding this plan of creation is that man's body is a mass of corruption—is earthy and earthly—while a certain portion of his functional or non-functional parts belongs to the supernatural category—that is, his body is natural and his spirit supernatural. All matter and force are therefore natural, except a certain entity called our spiritual nature. Expressed in other language, man is, unmistakably, the only animal who is permitted to evade the laws of nature, which all the brutes are obliged to obey. If this is not so, then to what natural laws is he obliged to render obedience? Which of them is he permitted to escape with impunity? Through this impregnation with supernatural power, he is endowed with an innate sense of good and evil, apprehended through divine revelation, or through obedience to the supernatural. Man is thus partly from above and partly from below, while all the beasts are exclusively from below. The solution of this dualism is of the weightiest concern in the solution of the social problem, which pertains to our moral and material wants, and if our spiritual needs form the basis, the longer the question remains unsolved the better. The reason why this dualism is solving itself is because economic man regards his material wants as the basis of his existence—at least his actions impel him in this direction, although his theories may run counter therewith —and so long as he devotes his energies to the gratification of his spiritual wants, he must certainly neglect his material necessities. All depends upon which he takes as the basis of his existence. He may get the one or the other, but he cannot get both. If he prefers the material, he must favour science; if he prefers the spiritual, he must favour abstraction. Economicism is conducive to our spiritual well-being; science to our material and moral. The material and the moral cannot be separated, and if morality consists in obedience to the laws of our being and development, the question arises, What further acts of morality does God demand? What actions inconsistent with these laws does He require us to perform? What interest or delight can He take in the physical extinction of His race of worshippers?

These speculations lead to constantly diverging trains of thought; for every new religion, or philosophy, is a tendency in this direction without destroying the older forms, ancient shapes of abstract thought being modified rather than de-

stroyed, and even what is called new is only so in a relative
sense.   In sympathy with the scientific method of inquiry, we
must proceed by observation, comparison, and analysis, not
taking it for granted that man possesses any instincts which
differ from those of the brute simply because our ideas lead us
into this belief.   The structural and consequent functional
analogies between man and the lower animals have already
been pointed out, and if there can be found anything fundamentally different in the moral or intellectual natures of man
and the ,brute, there will be strong presumptive evidence in
favour of the existence of a spiritual nature, but it would
not yet be logical to conclude that this spirituality is to be
attached to man.   Science can establish no other basis of
superiority than that of the powers of perpetuity in any
species of plant or animal.   If the brute can perpetuate and
develop itself by virtue of its own innate instincts, and if
man, by virtue of his education, can learn how to extinguish
himself, it seems rational to infer that spirituality pertains to
the brute, not to man.   By the abstract method of thinking,
however, structural extinction seems to favour spiritual superiority, whereas the scientist can see no conflict between structural, functional, and spiritual powers of perpetuity.   The
essence of perpetuity reposes in obedience to the laws of being
and development : the stricter the obedience, the greater the
powers of perpetuity.   Comparative anatomists have pointed
out that both absolutely and relatively, with insignificant
exceptions, man's brain is heavier than that of any of the
lower animals.   In examining the brain, or any other tissue,
two important factors must be considered : (1) the mass, and
(2) the texture or quality.   It does not follow that the maximum of force resides either in mass or texture separately, but
in an harmonious combination of both, coupled with the location of the organ.   Krahmer says : "The intelligence of the
brute manifests itself in the same way as that of man ; no
difference between instinct and reason exists except in degree."
Another famous researcher, Czolbe, says : "The impression
that animals are incapable of forming ideas, opinions, and
inferences is contradictory to the teachings of human experience."   The evidence of multitudes of investigators in

this field of inquiry can be educed to disprove numerous contradictory theories which have been advanced. The lower animals build houses, caves, nests, bridges, dams, &c. They hold deliberative assemblies, processes of criminal procedure, engage in migratory expeditions, and have languages which we cannot understand; they remember past events, and all their knowledge, like man's, is derived from hereditary predisposition and direct experience. Their sounds, signs, and gestures surpass those of many human tribes, and some of their senses are more acute than those of the most civilised of the human family. They have their feelings of love, hate, and revenge. It is by reflection rather than instinct that the fox, for example, provides methods of escape from his hiding-place, and steals fowls when he is least likely to be watched. It is experience that makes the older animals wiser than the younger. Crows and sparrows which are habitually shot at do not dread people without guns. The revengeful fights between swallows and sparrows are well known to ordinary observers, and the political and social communities of ants and bees are universally recognised facts. Ants, like human beings, wage war, fight battles, make expeditions, bring home slaves, and train them for their service, keep "cows" in stables, and practise agriculture. Who has not heard the marvellous docility of the elephant? The extraordinary sagacity of the ape, superior to that of some tribes of men, has been more than superficially investigated. Many gregarious animals choose leaders and obey their commands. Nor has development in the languages of the lower animals been neglected. Some human tribes, who indicate their feelings and sentiments by sounds, do so less perfectly than some so-called brutes. In man the sexual and self-preservative instincts, as well as those of the maternal love, is purely of the brute kind. The mother ape leads her young to water and washes its face, despite its crying, and wounds are washed out with water. When in distress the ape weeps like a human being, and in a manner which is very piteous and affecting. Many of the ape tribes construct huts or roofs, and sleep in a sort of bed. In battle they defend themselves with their fists and with long sticks, and in burying their dead they cover the bodies with leaves in a secluded place. African monkeys have been known to die of grief over the loss of their young. Emotions, entirely human in their manifestations, such as pleasure, pain, misery, have also been observed in the lower animals, and terror with them causes the muscles to tremble, the heart to palpitate, and the hair to stand on end. Courage, timidity, good and bad temper, rage, jealousy, desire for praise, &c., are also manifested. There are, moreover, exhibitions of curiosity and wonder, and the memory of some animals is surprising. Darwin's dog, for example, recognised him after an absence of five years. The powers of imagination are manifested in animals when they are dreaming, and the powers of reason are aptly illustrated by the way in which birds readily learn to avoid self-destruction by flying against telegraph-wires. The highest order of the ape cracks nuts with stones, and Abyssinian baboons attack their foes by rolling down stones upon them from the mountain-tops.

But all these actions, according to abstract theories, are quite natural, while man, by virtue of his religion, his education, and his civilisation, is a supernatural being. Is a civilisation which is doomed to extinction more supernatural than a barbarism which is destined to be perpetuated? A scientific civilisation, although, as we shall see, far removed from barbarism, can only be measured by its powers of perpetuity, and not by its liability to decay and to final extinction.

With all the evidence educed from this wide field of inquiry, the scientist is impelled to the conclusion that there is only one phenomenon—nature; and this is the only manner in which the dualism can be solved. There is no missing link between man and the lower animals; and, despite all his attempts to evolve into a civilised, an educated, an economic, a carnivorous, a cooking animal, and into a prince of scavengers, he has acquired no new instinct or faculty which removes him from the brutes. Admitting his higher capacities for evolving into higher stages of perfection, and for developing greater powers of perpetuity, yet his faculties have been employed in carving out a period of retrogression in his history. I will admit, however, that in the great epoch of universal evolution it may be marked as a period of progression; for I still entertain the hope that the written history of the past three thousand years, as well as that impressed upon the far more ancient geological pages of our earth, will be the means of enabling us to escape many a disaster in the epochs yet to be. When we begin to recognise nature in her process of evolution, our knowledge of her beginning and her end becomes logical sequences. Our efforts have been expended in vain attempts to force the unfittest of organic forms to survive, namely, those which have offered the greatest resistance to the laws of their development. The deeds of our ancestors are but stepping-stones to higher achievements.

Having thus reduced all conditions affecting our inquiry to one phenomenon, the way is paved for the evolution of wealth and morality. In the pre-organic period of our planet, all those materials and forces were wealth, which served to bring about those conditions favouring plant life ; and it was equally the destiny of the plant to subserve those conditions which brought animal life into being.[1] Animal life is distinctively characterised by sensation (subjective) and intelligence (objective), and these are the properties of nerve tissue, although these essences have a prior and an incipient existence in the blood. Age is a process of ossification—a transmutation from the organic to the inorganic. Wealth may therefore be defined as consisting of those materials and forces which conduce to the greatest quantity and highest quality of nerve tissue. In this definition no classification of animals is necessary, although I use the word man to signify the highest type of sensation and intelligence. Man's destiny is involved in the exercise of his structures and functions in obedience to the laws of his being and development. Wealth therefore consists of those materials and forces which are conducive to the fulfilment of man's destiny. In the widest sense of the word, wealth embraces man himself, for his own material and force are also exercised in promoting his development and consequent destiny.

[1] I have frequently used the expression "being and development," although, in reality, the word development expresses all that can be understood. A new form brought into being is merely the effect of prior developments. Nothing is actually created, although the word creation has a useful meaning.

CHAPTER II.

HAVING shown that all animals, including all their structures and functions, depend upon the soil for their being and subsistence—and this statement is directly or indirectly true, so that it makes no difference whether the animal is carnivorous, omnivorous, or herbivorous—it may be justly anticipated that land plays an important part in the solution of the social problem. Although land is the basis of the plant, yet this fact does not interfere with the conclusion I have arrived at, namely, that physical are subject to vital laws. Land is the basis of production, but vitality is the basis of perpetuity of production. The confusion of vital with mechanical laws— the organic with the inorganic — is the existing curse of science. This question is the chief corner-stone in our social structure, and no solution of the social problem can be made until economic forces are so changed that they can be made conformable with biological laws. I confine my remarks to land because those other equally important commodities, such as light and air, have not yet been monopolised, although they form a component part of scientific wealth. In its widest sense, land embraces all things which are conducive to the fulfilment of human destiny; in more restricted significations, it means the soil, and all its products —all monopolised commodities. The context will be a sufficient index to the sense. The land question is the parent of existing social conditions, and it is only through the same basis that a solution is possible. Land makes our bodies and sustains them : to deny us a title to land is to deny us the right to live—except upon charity. The heroes of our race are not those who delight to pawn their lives in charitable institutions. The right to land destroys the monopoly of our very souls. Must justice be denied that charity may abound? God forbid! So soon as we can assert and maintain the right to live, we destroy all monopoly and become free men. Then "liberty, equality, and fraternity" will reign supreme. Never has there existed a greater truth than that uttered by the anarchist when he maintains that political institutions only exist for the enforcement of monopoly. To vindicate the right to live is to transmute all economic rights into a figment, for it empowers us to draw on the property of others for our subsistence, or gives us the privilege to exert our labour upon such property; but the same law of right to live gives us equal title to death when the population grows out of ratio to the area of land required for its support. If this is not true, then the right of might must prevail.

The object of this chapter is to show the condition in which agriculture would now be found had scientific methods always prevailed. The drawing of the contrast between scientific and economic agriculture is necessary to an intelligent understanding of the subject, and to the devising of an efficient remedy.

Mankind has so gradually evolved into the present state of civilisation that no sharp line can be drawn between the man and the brute stages of our existence. The rapid decay of Christianity as a vital force—or, which is the same thing, the growing spirit of scientific inquiry—has given a great impetus to this inquiry, and there are many able investigators in the field; notably, Professor Dr. Büchner[1] and Abel Hovelacque,[2] Sir Charles Lyell, Dr. Karl Mayer, and other distinguished geologists have also shed much light upon the antiquity and habits of pre-historic man, while the reports of travellers amongst existing tribes have enabled us to draw useful comparisons. Dr. Mayer estimates the period which has elapsed from the middle of the Tertiary to the present

---

[1] *Thatsachen und Theorien aus dem naturwissenschaftlichen Leben der Gegenwart*, published by Allgemeiner Verein für Deutche Literatur, Berlin, 1887.

[2] *Les débuts de l'humanité: L'homme primitif contemporaire.* Paris, 1882.

time to be at least 250,000 years, which corresponds to the antiquity of man—that is, regarding us as being designated as man at that period in which our race contracted rude forms of our present habits, such as religion, family, trade. Historical man only dates back about three thousand years. During that remote pre-historic period, a time existed when, like many existing savages, man did not cover his nakedness, and had no shame to be hidden. Covering our bodies did not take its origin in any feeling of shame, but as a protection from weather and injury, and as our shame parts were the tenderest or most exposed, they were the first which received protection. In order to obtain security from the stings of insects, the whole body was sometimes covered, or rather rubbed, with obnoxiously swelling compounds, which discoloured the skin. Men lived in caves, or slept under leaves, and their condition at this period was not far removed from that of the apes; the latter, however, were more arboreal in their habits, their feet being more prehensile, so that they constructed kinds of nests on the trees. Much evidence has been advanced to prove that primitive man was an habitual vegetarian; but, owing probably to over-population, he was driven into scavengery, and ate fish, turtles, lizards, insects, carrion, &c., as well as vegetable foods. He even ate human flesh, and when the means of subsistence grew scanty, children were slaughtered for food, and old people were either killed or permitted to die of starvation, if not also consumed. The superfluous bodies of the dead were either buried or left lying in the open air. The remedy for over-population by means of war was not probably understood to any appreciable extent. Over-population must have been genuine, for there was no land monopoly in those days. Much charity could not have been expected, for it was impossible to accumulate property, there being no institutions to defend it. There was no idea of God or immortality, but simply a belief in good and bad spirits, and speech was very rudimentary. That state of fear, a species of which is possessed by some lower animals, developed into a sort of religion, and the devouring of human flesh became incorporated with the religious ceremonies. These religious characteristics adhere to some existing tribes.

The imagination, or idealisation of matter, developed a crude religion, and this, our original sin, is the source of those abstract ideas which form the basis of our civilisations. The scavenger life which our remote ancestors were obliged to follow detrimentally affected their moral qualities, and gave birth to physical degeneracy, and the development of the imagination was an obstacle in the way of observing the laws of nature. They did not adopt natural methods in their attempts to evolve into carnivorous animals ; hence the failure. Later on came the discovery of fire and cookery, by means of which the disagreeable flavours of our unnatural foods were destroyed, which aggravated their degradation, and the development of this practice has resulted in the cookery of foods which, uncooked, possess the most delicious flavours. This illustrates the immoral tendency adhering to the development of all abstract conceptions, for the development of all bad systems can only result in making them worse. By the invention of cookery, a larger number of unnatural foods could be utilised for consumption, and the vilest of filth in the animal world could be made sufficiently palatable for consumption. The pressure of population against the means of subsistence constantly intensified this mischief, and no remedy has yet been discovered. They then domesticated some species of the lower animals, using their carcasses for food ; and, when property became more secure, they branched out as agriculturists and monopolists. When land became scarce and valuable, it was monopolised by the priests, and the masses became slaves, which necessitated a strong government to protect the rights of property. In those days the priest was king and general, and the form of government was a theocracy, the right to rule and to monopolise being divine. These conditions are undergoing further development at the present day ; and, by way of contrast, let us now inquire what the condition of our ancestors would have been if they developed the scientific, instead of the abstract, parts of their consciousness.

Even at the present day, the theory prevails that, in new countries, the pioneers must have axes to hew down the forests, and implements and animals to cultivate the land.

This must be so, for it is the experience of practical farmers; and yet let us consider what the practice of thousands of years has taught us, and the same truths may be verified by the experience of a single generation of men. The removal of the forests has caused irregular rainfalls—alternating floods and droughts—the drying up of springs and rivulets, the development of insects injurious to vegetation, the loss of vitality in our agricultural plants, the scarcity of timber for domestic purposes, for the preservation of health, and for the protection of the country and of vegetation from fierce blasts, and many other calamities might be enumerated.

Economic practices, on the other hand, teach us that there is no money in trees so long as they are permitted to remain abundant—that they must be made scarce in order to increase the individual and the national wealth—and that their place must be occupied by domestic animals, because they bring money. There must be some solution of the dualism of scientific and economic practice. Both forms are the experiences of men, and why should the economic be taken as the basis of the scientific? Why should the practice of a generation or two of men be weightier than that of scores of generations? One of the most cogent reasons is simple, namely, that the greater experience, which includes the less, is profaned by the unconsecrated name of science, through which reason is made the basis of faith. In the abstract category, reason is of none effect save in so far as it rests on the foundation of our credulity. It is desirable to bear this dualism in mind, for my proposed institutions will be a disastrous failure unless they bring the greater and the lesser experiences into harmony.

Without denying that our remote ancestors were under the stern necessity of acting as they did, or that it is imperative for our race to undergo a process of abstract evolution, yet it remains to be observed that their methods were unscientific. Had their minds been drawn to observe nature, instead of indulging in flights of imagination and engaging in superstitious ceremonies, their methods of obtaining the means of subsistence would have been reversed. At this remote period agricultural plants and animals had to struggle for existence under the same law as the tribes which subsisted on them;

and the development of domestication and civilisation has
consisted in a gradual removal of this struggle. The sciences
of the present day are devoted to the devising of methods for
the further removal of organic life from the law of its development, and this tendency is specially observable in our
agricultural practices. Plants are fed with soluble fertilisers,
animals with easily digested foods, and tillage operations are
practised for the purpose of enabling atmospheric influences
to perform work which the roots of the plant ought to do,
such as the attacking of insoluble compounds in the soil.
Not only so, but domesticated life is protected from all kinds
of exposure, and receives the most careful attention and
cherishing. We have seen that labour applied to articles of
food tended to make them less valuable; we now also see
that labour applied to land makes the soil, in the long-run,
less productive, and it consequently also possesses less scientific
value. However, labour applied to land increases the economic value in two ways: (1) by making the fertility scarce,
and (2) by a temporary abundance of the products of the
soil, due to labour, and especially improvements in machinery,
an inflation of human desire is generated. Scientific wealth
thus suffers loss from two sources: (1) decrease of fertility,
and (2) decrease of vitality. With all these practical experiences, drawn from many centuries of observation and
research, we can now easily see the mistakes which were
made by our remote ancestors. Had they operated scientifically, they could have removed whole forests without the
use of any instrument, and they would have required no
implements or working animals for tillage purposes. Having
observed that they were not carnivorous, they would not have
brought any animals under domestication. There must have
been fruit and nut trees; and if the forest contained other
trees whose fruits were not serviceable as food, their plan
would have been the rearing of larger numbers of fruit and
nut bearing trees by planting young shoots in localities less
favourable for their growth than was their natural home, by
which method a few of the strongest saplings would survive,
and would ultimately displace the native trees of the forest.
This tendency could have been furthered by destroying the

vitality of the unserviceable trees through removal of the leaves and lighter branches; and in the performance of this work it must not be forgotten that arboreal man had five or six times the strength of civilised man. This is nature's method of converting forests into fruit gardens, and our ancestors need have performed no more labour than the birds do at the present day when weed-seeds are conveyed to remote districts, exterminating not only agricultural plants, but also varieties of weeds whose vitality has been lessened by subjection to agricultural operations. The same process should have been followed with reference to cereals or other articles of food, and by the same law they could have been so hardened in vitality as to exterminate the grasses, weeds, and other plants which were not used for human consumption.

Thus we find that labour has been the curse of our race, instead of a blessing, and any party who will succeed in founding new institutions on the basis of the aristocracy of labour will speedily suffer the consequences of their folly. It may be urged, however, that the scientific process is too slow, that the pressure of population against the means of subsistence necessitated the adoption of artificial agriculture, and that the work could have been more easily performed by the use of implements. The restriction of population will be discussed in a later chapter, but with reference to labour, it need only be said here that man himself would soon become extinct unless he had some method of exercising his structures and functions, and these exercises must have some relation to the methods for procuring the means of subsistence. Natural exercise, which differs very widely from our conceptions of labour, consists in gathering the fruits of the earth, not in producing them, but from this it does not follow that exercise should be confined within this limit. I confine my observations here to agricultural exercises, for, as we have seen, agriculture forms the basis of scientific institutions. It is now easy to understand how, under scientific institutions, the idea of property could never have originated. Why, even at the present day, a berry-picking excursion is regarded as an amusement, and so long as there are plenty berries for all, the iron hand of monopoly is fettered: indeed, the idea of such a monster could never have entered our wildest imagination. Another important deduction from our premises is, that the idea of trade could never have originated so far as the means of subsistence are concerned; for instead of people going after their food, the food would have come to them. Every family would have their necessaries growing at their own door. During this long period of 250,000 years, the tropical plants could have been made to flourish within the arctic circle, and *vice versâ*, and thus foreign, as well as local, trade could not have had inception, neither as an actuality nor as an abstract idea. There could not even have been a State for the enforcement of the rights of property or the punishment of crime; and any organisation that would have originated could only have been for purposes of co-operation in developing the means of subsistence. Here we see more plainly than in previous chapters the violence of the scio-economic dualism. By the economic method of thinking, it is not necessary to save our natural foods; for man, by virtue of his supernatural powers of objectivity, can utilise such plants and animals as he finds growing around him, and, by cookery and flavouring substances, he can make them palatable for people of the highest education and refinement. Besides, in the same train of thought, man, as a trading animal, can build ships and railways for importing and exporting foods which are restricted to certain localities, thus furnishing respectable employment for people of high social standing; and, by cultivating the soil, suitable employment for slaves is obtained, thus, by way of contrast, bringing out more prominently the virtues of education, wealth, good breeding, and the charms of civilisation. Moreover, by establishing a priesthood to monopolise the means of subsistence, respectable employment is obtained for professional gentlemen, especially politicians, and the theories of rights and duties are such as are best calculated to sharpen the cunning faculties of all who are permitted to walk within the social circle.

Having seen that man himself is part and parcel of nature, and must render obedience to natural laws, it is plain that he can only spread by virtue of the same laws which must be obeyed in the propagation of plants; and the lower animals can certainly not escape the same forces. The law of development is the same in all the varying forms of the vital world. It is extremely probable that man originated in a tropical climate, and his movements would, for the most part, have been towards colder regions. His ancestors, through millions of ages prior to this period, had no conception of clothing, and their only sources of heat were food-combustion within and the rays of the sun without. In his journeyings polewards, he was confronted with lower temperatures; and two other elements of warmth were consequently discovered, namely, fire and clothing, both of which are unnatural and must be consigned to the same category as the art of cookery. Clothing tends to destroy the functions of the skin in the same way as cooked foods tend to destroy the functions of our teeth, jaws, and internal organs, and can never take the place of the natural covering of hair with which the bodies of our ancestors, as well as those of many existing tribes, were blest. The bodies of many civilised people would still be densely coated with hair were it not partially worn off by the friction of their clothing. If clothing had never become fashionable, whether we were covered with hair or not, our bodies would now be as capable of resisting the cold as our faces. Man may evolve into a hairless animal, but into a clothes-wearing animal, *never*. When we consider the importance of the functions which the skin has to perform, and how it acts as a barometer of our health, we can clearly understand the injury inflicted by a deadening of these functions. And yet we are told by medical authorities that the wearing of clothes and the eating of hot foods are matters of economy in our dietary. Such practices may save food, but they waste land and vitality. Besides, if the hot foods are cooked, there is no saving even from the economic standpoint, for greater quantities must then be eaten. The evils arising from clothes-wearing, I admit, are to some extent mitigated by frequent bathing, which restores the activity of the skin, but this violent method, in the absence of the milder air, rain, and sun-baths, is often more injurious than beneficial. When we consider the subject in all its bearings, we cannot resist the conclusion that flesh-eating, clothes-wearing, alcohol-drinking, and food-cooking all bear a relation to each other, and the abandonment of any one of these practices should naturally be accompanied by the abandonment of all. It is a fortunate coincidence that civilised man has not evolved into a shame animal, and the idea is as much of an abstraction as the fiction that he is an economic animal. The wearing of clothes has been the source of more immorality and crime than all other habits combined. Had our ancestors not moved northward or southward more rapidly than they could have accommodated themselves to the lower temperatures, the clothes-wearing habit could not have originated; but this statement does not affect the occasional necessity for wearing certain garments or coverings under special circumstances, or in cold climates while the body is in a condition of repose. The injury inflicted is through the constant wearing of clothing under all temperatures, and a movement in the right direction would take place by enjoying an air-bath in open space for a few hours daily under more or less active exercise according to the temperature. These are not merely logical conclusions drawn from the science of development, but they have been amply proved by the practical experience of scientific vegetarians, who have felt astonished at the intense uneasiness and restlessness of their bodies while wearing the ordinary amount of clothing, due to the great activity of the skin, when they have confined themselves to scientific rations for several months. The desire for vigorous exercise or bathing in cold water becomes intense, providing the health and vigour have not been appreciably impaired by prior excesses, and the force of that electric vitality, which has been deadened for many centuries, reappears.

It must not be inferred from my remarks that there can be no difference in the application of methods between plants and animals in the struggles for existence, although the law is identical in both instances. Man can struggle scientifically without being obliged to injure or destroy his fellows, as he is now coerced to do, but the methods of attaining this end will be developed in the following chapters. We shall find, on the contrary, that the law of struggling for existence need not conflict with our highest ideas of human progress and sympathy.

When plants move from their home to struggle for existence in other localities, the weakest must perish; but with animals, none need move except those which have the strength to bear more unfavourable conditions. The law which man has violated is this, that he has migrated and at the same time attempted to carry with him his prior conditions. Instead of adapting himself to the climate, he attempts to make the climate adapt itself to him. All the grave errors of our ancestors we have developed with loyal fidelity; and what makes this truth all the more revolting is, that we do from choice those things which they did from necessity. We still consume the most disgusting species of carnivorous animals—the most loathsome scavengers—and they have become dainties for the *élite* of civilised society. If our ancestors devoured carrion from necessity, we, from choice, devour flesh tainted with all manner of diseases. What are we better than cannibals, when we consume animals which devour men? What we eat and drink is the best standard of our morality. Civilised man has failed to evolve into the king of the scavenger tribes, and there is no sense in carrying the experiment any further. If we could relate to our remote ancestors the various devices which we have discovered as remedies for over-population, they would feel appalled at what we are pleased to designate human progress. If we looked at them through their own spectacles, we would see that they were less barbarous than ourselves. If they could ask us what we have to show for the experience of so many centuries, what would our answer be? If they could ask us when we learnt to speculate with the souls of unborn generations, where would we go to hide our heads in shame? If man were a carnivorous animal, or even an omnivore, I would regard the fact as the strongest proof that there is no God, and that the only supreme being is the Devil.

CHAPTER III.

IN previous chapters I mainly made reference to land in the ordinary practical sense of the word, taking average crops as the basis of measurement, or leaving the reader to infer that all parcels of land have equal fertility and productiveness. It is now in order to discuss the scientific aspect, and here the same principles are involved as those which 1 have already discussed with reference to agricultural plants, namely, the chemical constituents, or the chemical analysis of the soil. We must ascertain the part which each chemical element or compound plays in the production of the plant on the same principles by which we have determined the functions of the plant compounds in the processes of nutrition ; but in the discussion of this question, the laws of vitality do not enter. Our foods are only known to the scientist by virtue of their protein, fats, carbo-hydrates, and salts, and when he speaks of land, he wants to know its capacity to produce these nutrients. In this reference land has a wide signification, for the carbon is derived from the atmosphere, the nitrogen [1] and salts being the only constituents derived from the soil. The nitrogen is derived from the decomposed organic matter in the soil, so that vegetable soils are rich in this element of fertility, and the salts are derived from the rocky fragments which, with the organic matter, constitute land in the limited sense of the word, and may be more definitely

---

[1] Small quantities of the nitrogen, however, are derived indirectly from the atmosphere in the forms of nitric acid and ammonia, which descend with the rains. With reference to the unsettled question as to what extent plants (notably the legumes) take in nitrogen from the atmosphere through their leaves, I may add that it is of little importance, the percentages being very small. The carbon is derived from the carbonic acid in the atmosphere, a poisonous compound which animals exhale, so that animals feed plants.

expressed by the word soil. Roughly speaking, land may be
divided into clayey, sandy, and vegetable soils, and various
subdivisions, notably calcareous soils, may be enumerated, all
of which have considerable scientific value, for these divisions
affect the mechanical texture and enable us to roughly determine the fertility and productive capacity. The salts which
enter into the composition of the plant are phosphoric acid,
potash, soda, magnesia, iron, lime, sulphuric acid, chlorine,
and silica. Sodium chloride (common salt) and silica are,
however, not absolutely essential for plant growth. These
mineral compounds are obtained by decomposition of rocks
through atmospheric action, and the roots of the plants also
attack the insoluble compounds and digest them. It is important here to bear in mind that the agricultural chemist
makes the same mistake in feeding the plant with soluble
fertilisers as the orthodox doctor makes when he recommends
easily digestible foods for animals; in both cases vitality is
impaired, it being a natural function of the plant to attack
insoluble compounds in the same manner as it is natural for
our vital organs to attack insoluble foods. This process is
the natural exercise for the organs of plants and animals,
without which the vitality becomes impaired, and in the
struggles for existence, the powers of perpetuity being enfeebled, other forms of life become fitter to survive.

As all soils do not possess the same percentages of the constituents of which they are composed, and as there is a good
deal of uniformity in the percentages which all plants take up
for their nourishment, we are confronted with a *law of minimums*—that is, the productive capacity of the soil is regulated
by the lowest constituent of fertility, just on the same principle
as the strength of a chain is measured by the weakest link.
An excess of the maximum constituents is often more injurious
to vegetation than beneficial. In applying fertilisers to soils,
it is only necessary to supply the minimum constituent, providing the others have equal degrees of availability for the
plant. To a limited extent, the same object is attained by a
rotation of crops, if those plants be selected which are less
exhaustive on the minimum constituent of fertility. It has
been ascertained by many thousands of experiments conducted

during the past half-century by able agricultural scientists, that the constituents which are more or less deficient in almost all soils are nitrogen, phosphoric acid, or potash, usually some one or two of these three. These constituents have therefore commercial value in the forms in what are commonly known under the name of artificial or chemical fertilisers.[1] The nitrogen is usually applied in the forms of nitrate of soda and sulphate of ammonia ; the phosphorus in the forms of bone-dust and phosphate rocks, usually made soluble by chemical processes, and the potash is usually supplied in the forms of sulphate and chloride of potassium, although wood ashes, which contain considerable percentages of potash, are also frequently used. The supply of nitrogen is to some extent under the control of man, for the removal of forests and tillage operations have a strong tendency to dissipate the organic matter in the soil, and it can be restored by reversing economic practices in agriculture. Potash exists in larger abundance than phosphorus, so that, in the most limited signification, land is reduced to phosphorus. We have thus analysed the pure gold of scientific wealth ; and, what is also of vital importance, we also find that all offences against phosphorus are scientific treason. Here we have the physical and the moral side of our inquiry, which cannot be separated, as we have seen in previous chapters. Where are our phosphate-mines ? The graves of the untold billions of our race, which are consecrated to the shades of our ancestors. The bones of animals contain 40 per cent. of phosphoric acid, as well as 53 to 54 per cent. of lime and small quantities of other fertilising substances, and yet this loss is trifling compared with the waste sustained in the excretions of human beings and other domesticated animals. Who can estimate the quantity of scientific wealth which is daily rolling down our rivers to

---

[1] Nitrogen has a higher economic value or price than the other two, potash being the least valuable ; but nitrogen is an element, while phosphoric acid (P_2O_5) is a compound of phosphorus and oxygen, the oxygen having no value, and potash is a compound of potassium and oxygen (K_2O). If these were quoted as elements instead of compounds—that is, phosphorus and potassium instead of phosphoric acid and potash—the apparently greater value of nitrogren would largely disappear, and there is no reason why the market quotations for the compounds, instead of the elements, should continue.

feed the scavengers of the sea? Such an estimate would be
a measure of the ill-gotten booty snatched from the battlefield in our triumphant victory over posterity. From the
principles just laid down we obtain another law, namely, that
the bodies, or even the excrements, of animals may be burnt
without destruction of scientific wealth, providing the ashes
be saved, for the organic matter which is dissipated into the
atmosphere—the nitrogenous and the carbonaceous compounds
—are not lost to humanity ; and even if they were, they do not
form the minimum element of fertility. And yet, forsooth,
all this scientific waste is an element in the production of
economic wealth ! The legitimate conclusion of this part of
our inquiry is, that our bodies should be cremated instead of
festering in our graveyards for the purpose of jeopardising
our health and furnishing savoury repasts for the Gods.

Socrates, just before his death, was asked what should be
done with his body, to which he replied in substance : "My
body; why, that is nothing: Socrates will instantly be no
more in the body, and it makes no difference what becomes
of the body when the spirit is gone." Had this great philosopher, however, made the following reply, namely : "Strew
my pinch of ashes upon the fruit-bearing earth to enquicken
a greater Socrates than myself: my barren dust in the sarco-
phagus is endless death "—had he made this answer, he would
have enunciated a doctrine whose scientific truth is not un-
surpassed by its spiritual loftiness, and he would have handed
down to posterity the key to all knowledge. This "heathen"
philosopher, however, was far in advance of our modern
professors of dogma, for they maintain that it does matter
what becomes of our bodies : they must occupy festering
graveyards and receive marbled immortality — they must
exhale pestilence in the nostrils of the living, and be kept
as reserve brimstone in the hands of his Satanic Majesty for
the purpose of meting out just rewards to all desecrators of
the shades of our great ancestors. If the scientist be permitted
to immortalise the body, he cannot consistently breathe a word
of disbelief in the immortality of the soul.

We are now in a position to understand what is meant by
the commodities of scientific wealth, which I have defined to

be "all materials and forces which are conducive to the fulfilment of our destiny." The soundest foundation is to regard every material and every force as being a wealth commodity, and then it must be subjected to two conditions: (1) the law of minimums, and (2) the manner in which the material or the force is utilised. Allow me to illustrate. In a previous chapter, I regarded, for example, potatoes, sugar, and butter as having no scientific value—and neither they have as articles of diet—but they may be converted into foods or other commodities which have scientific value. The sting of the venomous adder, if used to destroy human beings, could not be regarded as a wealth commodity, but its chemical compounds could be converted into agricultural plants. Fish have no value as food for herbivorous animals; but their bodies, when strewn over the land, make deliciously flavoured grains and fruits. Our planet contains a large percentage of clay—and pure clay does not enter into the composition of the plant—but it would be folly to attempt to grow plants on a heap of purely chemical fertilisers, so that clay is beneficial as a diluent and an absorbent of fertility, even omitting its other useful properties; and even animal manures, owing to their concentrated and soluble nature, would, by themselves, soon bring plant life into grief. Pure sand (silica) is also very abundant, and although the plant does not require it for food, yet its mechanical and other properties are as beneficial as those of clay. Think as we may, we cannot find a material thing or a force which, subject to the conditions above named, is not a wealth commodity. The air we breathe, although it is probably the most abundant of all agencies, if we perhaps except the light of the sun, is a commodity, but it would require an immensely increased quantity of phosphorus to produce human beings enough to breathe all the air. In like manner, the bodies of animals contain small quantities of salt, and yet millions of tons would probably remain after converting all the nitrogen, phosphorus, potassium, lime, &c., into animals and food for their support. Even bonds and mortgages are useful for the ashes they contain. We thus see that there are great extremes in the physical world, from phosphorus to air, and, in the moral world, there must be corresponding extremes in relation to offences. There exists nothing which is not, or may not become, a wealth commodity.

The mission of science is to demonstrate how the more abundant commodities can be best converted into those which are more scarce; and the first period of our development will be completed when all commodities will be so abundant that all desire to monopolise them will cease, which is the direct reverse of the economic policy. Then must also offences against commodities cease, for nobody will want to appropriate things which exist in quantities beyond the range of his desires. Two forces may work together in the attainment of this end: (1) the curtailment of desire, and (2) the increase of minimum commodities. We cannot change one element into another, *e.g.*, nitrogen into phosphorus, but we can prevent minimums from waste, and convert abundant commodities into those which will better enable us to fulfil our destiny. It remains here to be noted that the State, in so far as it exists for the protection of property and the administration of justice, must cease to exist, for its functions will be gone. From the Individualistic standpoint, the State will therefore become a nonentity. When there is no property to be protected, there can exist no crimes against property; and when each individual in every generation has equal opportunities for obtaining the means of subsistence, all desire for accumulation must cease. A necessary condition in the attainment of this object is the abolition of the right to labour,[1] for the enforcement of this right can only lead to the accumulation of property. The existing state of our social affairs may thus be traced to two sources: (1) monopoly, and (2) the right to labour. The latter, however, implies monopoly, for no right to labour for the accumulation of property can exist without first obtaining a monopoly. The abolition of all monopolies destroys the right to labour.

Having thus laid the foundation of scientific economics— and it should be borne in mind that my remarks have still

---

[1] Economic labour is here meant, which excludes the idea of exercise in gathering the fruits of the earth, or its equivalent in other occupations, and also excludes the idea of exercise for the development of our physical or mental powers.

been confined to agricultural labour, which is the basis, and which regulates all activity—let us now construct the moral fundament, which is inseparable from the economic. I do not here use the word moral in the ordinary dictionary sense; for I have nothing to do with abstract conceptions of morality : my search is for conditions, and if I have erred, I can only be criticised for not coining new words to express these conditions. Neither must morality be confused with virtue, which I shall discuss by-and-by : morality is purely objective, and virtue is exclusively subjective, so that these qualities belong to different categories. The moral man is he who obeys the laws of his being and development, and commits no offences against the means of subsistence. This dual condition may, however, be resolved into a monism ; for he who commits crimes against the means of subsistence prevents others, living or yet to live, from obeying the laws of their being and development—that is, the tendency of nature towards the production of the greatest number of the most highly organised animals becomes frustrated. Obedience to a law implies the non-interference with the rights of others to obey that law. In the ultimate analysis, morality is thus resolved into obedience. All disobedience can be ultimately traced to the relation between structure and function. Every structure has a natural function to perform, and all offences consist (1) in neglected exercise of function, and (2) in the performance of unnatural functions. All structures receive natural exercise while engaged in performing such operations and duties as enable the possessor to procure the means of subsistence. Under a gradual change of environments, there is a corresponding change of duties, causing a tendency in certain structures to disappear and in certain others to develop, and, there being also a prior and corresponding change of incentive, the natural law of evolution is obeyed. There must be an innate impulse in the performance of moral duties, and this impulse is virtue. These are the conditions of perpetuity, without which all animal life would become extinct. I intend no contradiction here between man and the lower animals when I speak of non-interference with the rights of others, for all rights imply corresponding responsibilities;

Q

this apparent inconsistency in the application of rights to the lower animals will disappear in the further development of our inquiry.

In contrast with this scientific basis of morality, let us review some of the immoral acts of economic man. The function of our teeth and jaws is to masticate our food, and we have attempted to perform the operation with our hands, so that, in the absence of natural incentives, we have gone through no process of evolution, and such a process would not have been conducive to the fulfilment of our destiny. Therefore all labour employed in grinding and cooking is immoral acts, because we have attempted to force certain structures to perform unnatural functions. The same offence has been committed in the consumption of animal foods, and the fact that we thereby draw on larger areas of land for our support, and cause enormous wastes of soil-fertility, are further acts of immorality, for we deprive others of rights which they should enjoy. Therefore all labour employed in the rearing and slaughtering of stock is immoral acts. As we can never evolve into inebriate animals, all labour devoted to the production of stimulating drinks is acts of immorality; so is also the wearing of clothes, for we attempt to make our clothing perform functions which belong to the skin and other organs. Sometimes our immoral acts retort mainly upon ourselves, and sometimes fall mainly upon others. For example, when a cook is engaged in cooking for several people, the immorality falls for the most part upon the cook, who must suffer the debilitating effects of the unnatural heat from the cooking-stove, although all equally participate in the pernicious effects of eating cooked and hot foods. All education connected with these practices is also immoral forces. Indeed, all education is immoral which does not develop our instincts. We are economic by education, not by instinct, just on the same principle as we have been educated to be shame animals, no such instinct ever having existed. We are also orthodox religionists by education, not by instinct. I refer, however, to religions which have been contracted under false conceptions of subjectivity. Man is by instinct a religious animal, but he can never evolve into an orthodox religionist. The basis of true religion is an instinct possessed by all animals. It would not be much of a figure of speech to say that all our education is immoral, for it has no subjective basis, and its tendency is to suppress our natural instincts, to remove us from contact with nature, and to make us artificial (spiritual?) animals. True religion is connected with virtue, not with morality.

When we now, however, consider this question from the standpoint of universal evolution, all these immoral acts form the basis of morality. It is unquestionable that man has been destined to pass through a stage of keen objectivity, and whether he is yet destined to pass through a further Socialistic stage of evolution before the dawn of true subjectivity need not be discussed here. Socialism, however, has so often proved a failure that no rational mind can see any necessity for repeating the experiment. The moral ground against it is overwhelming, namely, that virtue must emanate from the State instead of evolving from the inner consciousness of the individual. The idea is barbaric, and has no instinctive origin.

We have now arrived at conclusions which enable us to form a moral ratio, and to reduce all morality to mathematical rules. In words, the ratio may be expressed as follows :—

The most moral man is he who best fulfils his destiny and draws on the smallest area of land for his support.

Here we have what may be called a physio-ethic ratio in which a given area of land forms the antecedent term, and the various achievements of the individual form the consequent term. All that is now required is to reduce these terms to numbers. It must not be inferred that the principle involved is to compel the individual to draw on small areas of land—indeed, the reverse is, rather, the true conception, namely, to induce him to draw on the largest possible areas, providing he can show correspondingly increased results in the fulfilment of his destiny, but he will be prevented from drawing on large areas and yielding small results. It will be just as criminal for him to utilise too little land as too much, and the basis of his knowledge, and his morality, will consist in his being able to calculate closely the area of land which he requires to fulfil his destiny in the most efficient manner. It is plain that this question is closely allied with that of population, and when the time comes for scientific restriction, the most immoral men will be obliged to restrict their families, if they are permitted to have any, and thus none but the moral men will finally survive. In this manner, every law of nature is obeyed, not excepting the struggle for existence, and the law is carried out in a humane and intelligent way, instead of being left to the cruel processes which pertain to economic men and the brutes. In the following chapter, it is my purpose to manufacture the machinery requisite for the carrying out of this project; but before the ratio is further discussed one precaution is necessary, namely, the meaning attached to the word. land. In the very widest sense, when it includes the force of light derived from the sun, land is synonymous with wealth, and in this case it also includes minerals, coal, &c., in the bowels of the earth; but as it will not be necessary to use it in this enlarged meaning, it may be said that, in the widest sense, land includes the soil and the atmosphere; for no separation can be made, seeing that about 90 per cent. of the foods we eat come from the latter source, leaving only 10 per cent. to be drawn from the soil. In a more limited sense, the meaning is synonymous with soil, and in the most restricted signification, it means phosphorus—the minimum element of soil-fertility. The context will usually be sufficient index to the meaning, and as I apprehend no confusion, I may be spared the alternative of coining new words to express these various meanings. As already pointed out, the individual has a title to the area of land required for his support.

CHAPTER IV.

ORIGIN AND DEVELOPMENT OF THE SCIENTIFIC STATE.

HAVING shown that the land, of necessity as well as axiomatically, belongs to humanity, it is evident that any given generation can only act as executors or administrators for posterity. No human laws enacted with respect to land titles can change or repeal those written on the face of nature. If the individual worked his plot of land, if it were set off by metes and bounds, and if the population remained stationary, each could carry out the will of posterity without any necessity for social organisation; but this basis is only theoretic, for although each individual is supposed to occupy and work his plot—and it must be borne in mind that he cannot be deprived of this area, for the fact of his being alive forces him to draw on his parcel for his support, and to deprive him of his land means that he has no right to live—yet this theoretic standpoint does not interfere with his right to engage others to work his plot or to exchange its products, in whole or in part, for those of any other parcel of land in any part of the world, so that in practice there need be little change in existing customs: some may work small gardens and others hundreds of acres. What is required to be known is merely the area of land, measured by its products, which each individual requires for his support. Without a knowledge of this detail, the population could not be adjusted to the area. It now stands to reason that a social organisation is required; and when we also consider the consequent term of the ratio, the necessity for such is still more conspicuous. The material and the moral being inseparable, it is plain that the functions of organised society are not only to determine the area of land which each individual draws on for his support, but also to measure his moral achievements—

that is, to determine the ratio between the plot of land and the destiny of the individual. The person stands between his plot and his destiny—between the matter and the force which it exerts. The social organisation which enforces this principle is the scientific State. Its function is to carry out the laws of nature, no other laws being necessary, human or divine. It is evident that the basis of this State is purely and absolutely individualistic. The individual must determine for himself whether or not he is to go down to posterity—to enter the kingdom of Heaven—and he must, therefore, have absolute control over the area of land required for his support. The State cannot dictate what he shall eat, drink, or wear; its function is merely to determine his physio-ethic ratio, and the law of social restraint, or scientific fashionism, will be the administrator of justice. It may be said that this is not an individualistic basis, that Individualism means the performance of an immoral action and the shifting of the responsibility on others. All I have to say is, that nature knows nothing about economic Individualism.

In order to understand more clearly what is meant by reducing our physio-ethic ratios to numbers, let us take an illustration. A holds the theory that man is an omnivorous animal, and that he can be healthier, stronger, and happier by keeping himself comfortably warm ; while B maintains that man is a vegetarian, and that health, strength, and happiness consist in keeping oneself comfortably cool. The result will be that A will require, say, four acres of land for his support, while one acre will be sufficient for B. Now, the State, in determining the physical and mental achievements of these competitors in the struggle for existence, finds that A obtains, for example, say, 800 of such numbers as may for the present be called units of value. His ratio will therefore be 4 : 800, or 1 : 200. If B now obtains the same number of units, his ratio will be 1 : 800. In this case the moral achievements of both competitors are identical, but B has performed his duties on one-fourth of the quantity of land which A has drawn on, and is therefore a more moral man. Now, it is evident that A, if he is ambitious enough to struggle for existence or found a family, will be compelled either to increase his units or reduce his area of land, or both.   In order to obtain B's standard of morality without reducing his habits of living, A must obtain 3200 units, making a ratio of 4:3200, or 1:800; or by reducing his land to two acres and increasing his units to 1600, he can accomplish the same results.   If B now attempts to obtain the same ratio on, say, half an acre and fails, his ratio now being, for example, 1:200—that is, equal to A's—both competitors are equally immoral, for they have equally miscalculated the proper area of land required to fulfil their destiny in the most efficient manner.   It is as immoral to draw on too little land as on too much.   From this basis all the functions of the State may be derived; and it is my purpose in this chapter to show how the scheme may be made practicable.   It is further to be observed that the basis of the State is a social compact.   The conditions of this compact may be reduced to one clause, namely: "When population begins to press against the means of subsistence, we hereby agree that only the most moral people shall be permitted to live in posterity."   When a majority of the people in any generation agree to this document, the social problem will be solved, and the devil with all his hosts will be routed.

Economic forces are now reversed, and we find ourselves in the scientific or Humanitistic category.   The supreme object now is to manufacture human beings instead of sacrificing them on the altar of trade, and the officers of State are a business corporation elected by posterity for the purpose of carrying out its supreme will.   Agriculture, manufacture, and trade must all be made subservient to man's destiny; and as the aims of the State and those of the individual are identical, no collision of interests can occur.   The method of election of State officers must also be the direct reverse of the economic system.   There can be no elections in the economic sense of the word.   The most moral men, by virtue of their morality, are permitted to become officers, the most moral men having the first choice; but the acceptance of an office becomes a duty rather than an obligation.   All positions, whether they belong to the State or not, are open to the most efficient or moral competitors in the respective departments, so that no idea of elections can be conceived, and the officer retains his position so long as he keeps in advance of all competitors.

Society now being a co-operative enterprise for the manufacture of machinery called man, it is necessary to analyse this machine in order that the greatest quantity and highest quality of goods may be obtained, and it should be borne in mind that the structures and functions of animals do not differ in principle from those of inanimate machines. Animals may be classified by their temperaments, but these are less pointedly marked in man than in lower animals, owing to the abstract character of his education and the artificial nature of his habits of life. These temperaments or systems exist in every individual, but there is usually a predominance of one or two of the three divisions; sometimes, however, there exists what is called a balancing of the temperaments—that is, there is little or no visible difference with reference to their predominating characteristics.

1. The *mental* system, or subjective temperament, implies a predominance of brain and other nerve tissue, and is the source through which thought and sensation are manifested. When this system predominates, the possessor has a sparkling, intelligent eye, a clear, sharp voice, is extremely sensitive to enjoyment, pain, and suffering, is quick in movements, highly refined, clear-headed, prone to study, has fine, light hair, small bone and muscle, a soft, delicate skin, and a highly magnetic touch; but being deficient in vital and motive powers, vitality is easily exhausted.

2. The *motive* system, or objective temperament, is characterised by highly developed bone and muscle, giving the possessor great powers of endurance and great tenacity to life; he is full of bodily activity, uses sledge-hammer oratory in preference to refined language with its accompanying gracefulness of manner, is a great pioneer and plodder, has strongly marked features, a dark complexion, a compact frame, a strong determination, and is an accomplished schemer, although the mind operates slowly owing to the deficiency of quality of nerve tissue.

3. The *vital* system, which embraces the digestive organs, the heart, the lungs, and the viscera, is marked by large

lungs, a strong circulatory process, large digestive and assimilating organs, a profusion of blood, and strong animal spirits. Proper breathing and an abundant supply of fresh air are necessary for the full development of the vital structures and functions. When this system predominates, the possessor readily accumulates an excess of fatty tissue, and there is not sufficient mental or motive incentive to work it off. He enjoys easy, out-door exercise, dislikes study, is very impulsive, enjoys life, is inclined to be shrewd, as he acquires knowledge more by conversation and observation than by study, is very showy, and makes an excellent priest or politician.

In well-balanced temperaments there is a blending of all the qualities and dispositions named, and various subdivisions and combinations could be enumerated if space permitted. The scientific and the artistic temperament cannot be separated, for in both instances there exists a keen desire to pry into nature, and as the natural instincts predominate, there is an abhorrence of educational rules, methods, and restrictions. The natural desire is to follow one's bent, and to ridicule the follies of the multitude. The possessors of this temperament are not money-making animals; they would rather starve, although this preference often leads them to fame and fortune. Too frequently, however, this brilliant temperament is perverted by social intercourse, and the possessor is then educated to act as if he were an objectivist. There are many indications by which the person of education may be distinguished from the person of instinct; and I have taken special pains in observing that the former has a preference for whelps, especially such carnivorous animals as the dog and the cat, while it is the joy of the latter to kiss lambs and foals. We might consider many tribes and nationalities, past and present, and classify them in relation to their temperaments as affecting their subjective and objective traits; but the influences of education have been so potent that peoples have been compelled to act contrary to their natural impulses, so that an inquiry of this kind might be misleading in many respects. Although there is a close and unmistakable relation between all our structures and functions, yet I have little faith in the progress of physiology, phrenology, physiognomy, and kindred

sciences so long as the subjects are mere economic animals—
so long as we attempt to force structures to perform unnatural
functions. Our heads and faces cannot be successfully examined
so long as our bodies are inaccessible, for the part should be
studied in connection with the whole. When a law is not
distinctly stamped on the part, it may be learnt through the
general law of the whole. Even after our decease, superstition practically forbids access to the organs of our interior.

We must have a sound body before we can have a sound
mind—a sound structure before we can have a sound function. The scientist wants the function by all means, but he
takes the structure as the basis; the failure of the abstractionist is that he wants to save the function, and as for the
structure, it makes no difference about that. It were devoutly to be wished that he attempted to carry out this
theory in matters pertaining to Church and State.

A human being is most aptly compared with a steam-
engine. An engine must have a good vital system—that is,
a furnace sufficiently spacious to produce heat proportionate
to the energy to be expended, and the fuel must conform to
the size and structure of the furnace. The motive system,
or boiler, must be strong enough to stand the pressure; and
the nervous system, or engine, must possess intensity or
massiveness proportionate to the velocity or ponderosity of
the machinery to be set in motion. If velocity is required,
we must have a slimly constructed engine with driving-wheels
of large circumference and with piston-rods of long stroke—
just as the man of velocity has long limbs and sharp perceiving faculties. If the engine is for heavy machinery, moving
at a slow speed, quite a different construction is required. This
mechanical law pervades all animals, and is aptly illustrated
by the nimbleness of the weasel and the sluggishness of the
sloth. The structure is an unfailing measure of the function,
and our inability to make close observations and draw logical
conclusions does alter the principle. When we once recognise
the law, we must conclude that it operates under conditions
beyond the reach of our faculties. The law cannot strictly be
expected to apply to domesticated animals, however, for we
manufacture them for our own use, while nature manufactures

them for their own good; and we must aid nature to make ourselves for our own good, and not for the good of trade. In speaking of the relation between structure and function, general rules cannot make allowance for flaws that may occasionally appear in the material. A boiler made from a certain brand of plate is calculated to stand a certain pressure; but in practice a test by hydraulic pressure is usually made for security against flaws. So it is with pieces of timber: a given species of wood, grown under given conditions, will stand a given test of strength compared with other species, but from this it does not follow that all pieces of timber from the strongest species are stronger than some pieces from the weakest species. The perfection of material is when the brand is an absolute guarantee of strength, no special test being necessary. So it is with animal mechanics: every bone, muscle, sinew, and organ has a certain strain to bear, and the work which every species of animal is decreed to perform is connected with the struggle for existence. In man, however, this law is constantly violated; for we find in many savage tribes a disposition to deform the structure out of conformity with the natural function. Hence we find wedge-shaped heads, thus deformed by artificial means; also flat and round heads, small feet, &c. There was an ancient tribe which practised the slitting of their tongues. If the intensity of barbarism were measured by the amount of injury sustained by such practices, and their consequent irrationality, I would instance a race of barbarians whose fair sex flatten their waists by tight lacing; but I purposely refrain from making any allusion, because I do not wish to trifle with people's credulity. Such methods of deformity may be traced to very ancient customs or religious superstitions.

The plain duty of the individual now is to exercise his mental, motive, and vital structures; the maximum degree of development is attained when they are exercised to the fullest capacity without strain, providing the proper quantity and quality of nutriment be supplied. All other conditions being equal, the most moral man will be he who attains this degree of perfection; but if some men gain an advantage through hereditary forces, the morality is none the less meri-

torious. The antecedent term of the physio-ethic ratio is
essentially an agricultural, and the consequent term a mechanical, inquiry. The departments of State will therefore be
primarily of two classes; and the duties of the officers will be
to establish comparative standards, with the view of ascertaining which individuals fulfil their destiny most efficiently.

Starting with the theoretic basis that each individual is
tenant for life of the plot of land which he draws on for his
support, we are in a position to evolve our system of agriculture. We may suppose, at the outset, that each occupies and
works [1] the plot which he determines upon for his subsistence;
but should some individuals choose other occupations, or should
A desire to exchange products with B, a contract is founded
between individuals with which the State has nothing to do.
If A does not till his plot, the work must be performed by
B (or B and C); and if B consents to till A's plot, A must
certainly render an equivalent to B. Although A does not
work his plot, he cannot be deprived of its products; and if
B is foolish enough to do A's work without receiving an
ample equivalent, he loses a yard or two in the race for
existence. If C's plot is specially adapted for wheat and D's
for oats, an exchange of these products will be of advantage
to both; for the area of land which each requires for his
support is thereby reduced, the conclusion being that an immense impetus will be given to trade, and that it will be
impossible to force trade out of its natural channels. The
idea may be entertained that this scheme, carried out amongst
so many individuals, will lead to great complications, but the
reverse is really the case; for it simply means that every individual may go to the market and purchase all his needs,
but instead of paying money, which system of exchange is
abolished and no money is required for any purpose whatever,
he must know the land equivalent of each commodity he pur-

---

[1] I use the word "work" with a special meaning. I have used the
word "exercise" to mean the gathering of the natural fruits of the earth;
but even should society resolve to return to natural methods, it will
require many centuries to attain this end, so that, meanwhile, the land
must be "worked." I confine the word "labour" to the working of
plots which we do not need for our subsistence, the privilege to labour
upon them being, therefore, a monopoly, so that labour is a criminal
offence.

-chases, and here some decimal scale of calculation must be adopted for the sake of simplicity. This process can be made so simple that all commodities may be reduced to land equivalents by an easy effort of the mind—mental arithmetic. The transfers are made by book accounts, on the principle of the existing system of credits, but there is no such thing as getting into debt. Such a thing is a physical impossibility in the scientific State. Each person's account, therefore, shows the area of land which he has drawn on for his subsistence,— and a decimal system will also be necessary for the measurement of land areas. For a basis of this sort, the acre is much too large an area; the one-hundredth part would be nearer the mark.

It is further evident that each individual must occupy some portion of land, even if it is only the spot which his house stands on. Linking this with the fact that agriculture is the basis of knowledge, that everybody must know how to connect himself with the means of his subsistence, and that his success in the struggle for existence depends upon the practical and scientific soundness of this knowledge, we arrive at the conclusion that some small portion of land must be connected with every residence, and tended by the occupier. I refer here to parcels of land which are less in area than those required to be drawn on for the full means of subsistence, so that the larger parcels will be greater than the areas necessary for the support of the individuals or families living in the residence connected with the portion of land on which it is built. For this purpose, three classifications of land parcels are necessary. 1. The smallest parcel of land connected with any residence must be sufficiently large to utilise as fertilisers all the excretions and other waste products of the household. This involves the minimum of agricultural work, and does not interfere with regular occupations of the members of the household ; on the contrary, it is less than the amount of out-door employment necessary for the preservation of health. 2. The second class has a different basis, and the size of the parcel is regulated by the quantity of garden products— fruits, vegetables, and nuts—required for the full supply of the family occupying the residence. 3. The third class

parcels, in addition to the above-named products, are so extended as to be also suited for the production of grains,
meat, and dairy products. I here include stock and dairying, because I have no right to presume that, in an individualistic State, all the inhabitants will adopt a vegetarian
diet. The main reason for separating the first and second
from the third class is, that fruits and vegetables may be
conveniently grown on small plots of land with a minimum
of work and outlay, while the products of class three demand
the expenditure of work and machinery, and are therefore
not adapted to small parcels of land. Agricultural progress
does not consist in knowing how to work the land and raise
heavy crops, but to know how to leave the crops alone and
let them raise themselves, the ultimate consequence being
the evolution of work into exercise—that is, the reduction
of human efforts to the gathering and storing of the fruits of
the earth. The necessary preparation for this process is the
establishment of centres where agricultural plants grow wild,
whence they may be brought to the occupied land and undergo a slight operation of tillage until they are able to take
care of themselves; but a certain amount of pruning will
always be necessary. For the attainment of this end, the
wretched idea that plants are improved under domestication
must be abandoned; the direct reverse is the case, for
domestication has the effect of diminishing both the flavour
and nutritive properties of all foods, and domesticated
animals may be placed in the same category in this respect.
Loss of nutriment, loss of flavour, loss of vitality, and increase
of propensity to disease all go hand in hand, while these
properties are reversed by increasing the hardiness of plants
and animals—that is, gradually exposing them to less favourable conditions than those which exist in their native abodes.
By this system of agriculture no fertility is wasted, and cities,
which are an economic idea, cease to exist. The process of
countrifying the city and citifying the country gradually
takes place, for when residences in towns and cities become
old and tumble down, the plots on which they stand are
converted into gardens, and the new residences are built in
the suburbs or in the country. The ultimate of this process

of evolution will be that the whole country becomes one city—or, expressed in other language, all the cities and towns become one country place. Those who like cities may call England, for example, a city, and those who prefer to live in the country may name England a country place without a city, a town, or a village. All the difference is in the name and in one's method of thinking.

Before entering into the question of agricultural education, let us discuss the mechanical department of the State. Agriculture does not exclude the mechanical idea, for it embraces physical exercise, although the primary conception is the intellectual development connected with the procuring and the perpetuating of the means of subsistence. It is the sum of all the sciences, and is the regulator of all other pursuits. Going back again to first principles, A, who does not work his entire plot, must render equivalent service to B, who performs A's agricultural duties, and A's work must therefore be equivalent to B's. At the outset B will work, say, three hours a day for A, so that A's work for B will be the expenditure of a corresponding quantity of energy. With the development of agriculture, however, B will be able to perform A's work in, say, one hour per day, so that A's work for B will be correspondingly deduced. We thus deduce the law that agriculture is the regulator of all work, and when no work is required for the production of food there will be nothing left but exercise. Labour cannot exist without monopoly and consequent robbery, which leads to the accumulation of property for ourselves and our "heirs and assigns for ever." All property is robbery ; and the full significance of this eternal truth is only half understood. He who labours on other men's plots without their consent and appropriates the products is a monopolist, a robber, and a plunderer ; and not only so, but he also deprives others of the privilege of exercising their functions in the natural way, thus bereaving them of their vitality as well as of their property. The fact that the rightful tenants of certain plots are not yet born is no excuse for plundering them and robbing them of " liberty, equality, and fraternity." Now, in order to understand the relation between A and B more thoroughly, we must know what their duties are, which consist in the exercise of their nervous, motive, and vital functions in such a way as will enable them to obtain the means of subsistence by the most approved methods of the age in which they live. If each works his own plot and has no transactions with the other, their duties and responsibilities are plain ; but if their plots are not ample for the support of themselves and their families, a very insignificant portion of their time would be employed in gathering the natural productions of their plots, and if they went idle for the remaining portion of the time, they would ultimately rust out and cease to exist. Another law comes into force, however, which prevents this calamity, namely, the families grow out of proportion to the size of the plots, and, there being no other land within reach, the two families engage in constant warfare against each other, and the stronger members keep destroying the weaker until the population is reduced within the means of subsistence. This law of exercise of function pertains to all animals, including man ; but man, by virtue of his superior intelligence, has introduced another law which the lower animals have not yet, as a rule, discovered. Instead of killing off all the surplus population by bellicose methods, he introduces an artificial system of agriculture, which sometimes temporarily increases the means of subsistence, although ultimately frustrating all his designs, but has the advantage, through the creation of labour, of preventing him from rusting out—or, rather, forces some to rust out and others to wear out. The scientific problem now to be solved is how A and B can restrict their families within the productive capacity of their plots, live neighbourly lives, and be prevented from rusting out. The law of the restriction of families will be discussed by-and-by, and I shall here proceed to inquire if any incentive can be found to induce men to exercise their nervous, motive, and vital functions under conditions in which the means of subsistence are abundant. I have already shown that neither A nor B can be deprived of the products of their plots, and that no amount of idleness or indifference on the part of either can deprive him of the means of subsistence ; the right to live carries the right to draw on the products of one's plot. But the right to live in the world implies the right to come into the world, and the right to come into the world implies that there is room there. If not, it implies the right to elbow somebody else out of the world. Nature's law of development implies the right of the stronger to come into the world in order to drive out the weaker, and no powers on earth can change the principle involved in this law, although its stringency may be mitigated. If twenty-five animals inhabit a territory where there is only food for twenty, it is plain that the food cannot be equally distributed, for the thereby engendered deterioration of the species would end in its extinction; if five of the strongest were removed the same results would follow, so that the preservation of the species can only be secured by the constant weeding out of the weaker members. Stagnation, produced by weeding out the medium individuals or other causes, will end as fatally as retrogression, but a longer time will be required; for so long as there are other species of animals which progress by constant removal of the weaker individuals, they will ultimately win the race in the struggle for existence. If A and B now agree upon an armistice, substituting for hostilities a contract for the restriction of their families within the compass of subsistence, the terms of the contract will decide whether A and B are jointly or separately responsible for new claimants upon the means of subsistence; but it is evident that the contract cannot be made in violation of natural laws, for it would give rise to future complications, and if the laws of nature are obeyed there can arise no necessity for a contract—except in so far as the meaning of the word is so limited that it only implies an agreement to obey the laws of nature. Any social compact which may be interpreted to have a different meaning must lead to anarchy.

The individual now being at liberty to purchase all his needs on the markets, the question arises, Where are the inducements to manufacture the commodities and bring them upon the markets? Having abolished labour, which is unnatural and criminal, I have also abolished wages. People demand wages for two reasons: (1) because they dislike labour, and (2) because it is the means of procuring subsistence. What I now propose to do is to reward people for their exercise. A large number of people exercise their functions without any hope of reward, save to be strong and have good health, and if in addition to this I also give them scientific wages, I am sure they will be highly pleased. By abolishing what people do not like, and rewarding them for what they willingly do without reward, I think I should win the sympathy and support of every member of the community. There is no alternative to this conclusion, even should I not feel disposed to be so liberal, for economic methods must be reversed in every particular. It may be said, however, that there still exists such a thing as work as well as exercise, and that people will demand wages for their work; to which I reply that the adherence to certain economic rules during the transition to the scientific system is no fault of this system. However, I will suppose, for the sake of argument, that people, meanwhile, demand pay for their work, and I will now show how they can be paid without the use of money, even after the organisation of the scientific State. In order to effect this result, I must here explain what is meant by units of value, referred to at the beginning of this chapter. It makes little difference, except as a matter of convenience, what is taken as a unit of value, although I shall discuss this question presently; the essential character of this standard consists in the bringing of all commodities or values into a just relation with the established unit, and for this purpose let us take the average time and energy expended in the performance of a given piece of work. For the expression of this unit I require a new word—*chronop.*[1] Still confining my remarks to agricultural operations, experiments will be required with reference to the time occupied in the performance of each operation under average conditions in the different classes of soil, and accurate estimates as to the yields per acre must be

---

[1] Greek, *chronos*, time, and Latin, *ops*, work. I must here wage war with the philologist, who will maintain (no better word being within reach) that I have no right to drop the *s* in *ops*, but I do so for the sake of euphony. I contend that it is better to butcher the dead language than the living. If I retained the *s*, the plural would be *chronopses* instead of *chronops*, and the adjective would be *chronopsic* instead of *chronopic*. Moreover, I contend that the accent may rest on the first or second syllable, or both, according to whether the time or the work is to be emphasised.

made for the purpose of establishing general averages. The time and labour expended on each crop now being known, the tiller receives credit in chronops, not for the actual time and labour which he expends, but on the basis of the general averages under his particular conditions; so that if A and B accomplish the same results with the expenditure of different momentums, A expending twice as much time and force as B, their chronopic credits remain the same, and B will have the further advantage of saving time, which he may devote to the earning of chronops in other departments pertaining to his destiny, the tendency thus being towards the raising of crops without tillage operations.

We now see clearly and distinctly from two standpoints that the farmer is not entitled to the full product of his work, if he produces more than is required for his own subsistence and that of his family : (1) because, as we have seen, each individual in the State is entitled to draw on the products of the area which he selects for his subsistence, and (2) the farmer is paid in chronops for all his work. What is his inducement to work for chronops? The answer is plain, namely, that they, in connection with the area of land which he draws on for his support, enter into the formation of his physio-ethic ratio, and thus decide whether or not he is to be promoted and to live in posterity. Plainly speaking, he must struggle for existence or cease to exist. It is further evident that no farmer who tills more land than he requires for his own subsistence will do so unless he receives chronopic equivalents. Here we have the basis for all other occupations, and this scheme will have the tendency to force all individuals into those employments in which they earn the most chronops, just as trade will be forced into its natural channels. The same principles and incentives apply to all branches of industry, with the result that people must struggle after morality, instead of money; and the most immoral people will be weeded out, instead of the most moral. Under economic conditions, the most moral people, detesting the tricks of trade, have no incentive to struggle after money, and they are therefore weeded out, because they cannot, as a rule, afford to get married and raise families. Many who are not removed in this way fall on our fields of battle. Under the plan just described, only such commodities will be manufactured which will enable people to exercise their functions, and exchanges will be made in chronops, just as under the present system of transferring transactions by accounts instead of paying cash.

The development of this system is an important feature. It is plain that people cannot get all the work they desire, and that an even distribution of the work cannot be made. The amount of work will always be growing less, and the desire for it will be correspondingly on the increase, so that if the system were confined to work people would finally rust out. But it does not follow that people must cease to exercise their functions when their poverty-stricken task of work is done. People can exercise their functions just as well when there is no work whatever to be performed, providing always that there exists sufficient incentive. For the purpose of evading this difficulty, all distinction between work and play must be abandoned, and every individual must receive the same chronopic credits whether he exercises his functions in the field, in the factory, in the gymnasium, in the studio, in the museum, or in the rowing-boat. By the adoption of this plan there can be no limit to the opportunities for earning chronops. If some people would rather play than work, the credits can be so arranged that more chronops can be earned by working than playing; but there is no danger of clashing here, for the amount of work will be so small compared with other forms of exercise that it will come to be regarded as play. The boy does not imagine that he is working when he is building a play-house; neither does the little girl when she is making dresses for her doll. Her only complaint is that she has not dolls enough to be dressed. "But some kinds of work are dirty," exclaims the critic. I challenge anybody to point out any dirty work, except when a structure is performing a false function. Reduced to its final analysis, my system merely forces every individual to exercise his structures in the performance of their natural functions; and when this end is attained we enjoy "liberty, equality, and fraternity" in the fullest and deepest sense of the phrase. All exercise develops into labour—and very dirty labour too—when the structure performs false functions.   Idleness is a false function of all our structures.   Indeed, both work and labour vanish when all structures perform their natural and legitimate functions; but when we attain this degree of perfection we become Gods, having no desires.

From what has just been said, the functions of the State can be easily evolved.   As already observed, there are two main divisions : (1) the vital (agricultural), and (2) the physical (mechanical).   The State has no control over the individual in anything pertaining to the exercise of his functions.   He who is obliged to suffer the consequences of his conduct must have unlimited control over his own actions.   There is only one principle on which the State can act.   As we have seen, society has agreed that the most moral individuals shall hold the most honourable and responsible positions in the respective departments, whether within or without the official circle, and it is therefore clearly the right and duty of the State to have full control over such affairs as will enable it to determine who the most moral people are.   In a word, the State is merely a body of examiners.   This fundamental principle, however, does not interfere with other duties which society may think fit to hand over to the State from time to time, and again withdraw at their good pleasure.   The State, as an organised body, may have the machinery for undertaking many duties which naturally belong to the individuals, and an individual may be an officer of the State in a very limited capacity, his main duties being bound up with the unofficered classes.   It is the most imperative duty of the State to arrange the chronops in such a manner that the individuals will be induced to exercise such functions as will enable them to procure the means of subsistence in the most scientific way known to that generation.   Otherwise expressed, the structures which are brought into the most active use during work must have corresponding importance during play.   All sports, games, amusements, &c., must therefore be conducted on scientific principles, and the performers must also know what portion (or portions) of their bodies is brought into the most active play with every movement of the body.   The physical and mental training must therefore go hand in hand, and every movement of the body can be ultimately traced to its effect in procuring the means of subsistence. These duties belong to the mechanical department of State, and they again form the basis of all the mechanical sciences, just as agriculture is fundamentally a biological pursuit. The individual, as we have seen, has three classes of structures to be exercised, namely, the nervous, the motive, and the vital system, and the chronopic units must be so adjusted that none of these systems be allowed to degenerate, but that all receive a sufficient amount of exercise to develop them harmoniously. Matter being the basis of force—structure the basis of function—all the structures of the body must be examined and justly chronopised. For example, weak eyes are usually caused by excessive indulgence in midnight studies or other employments, so that, for the purpose of breaking down these ruinous practices, a high chronopic value could be attached to sound eyes, which would have a tendency to compel people to rise early in the morning, instead of sleeping by day and seeking their prey by night, and the area of land required to furnish the light would be saved. This rule would also have special weight when applied to our vital system, for the man who attempted to live on cooked food, and then exercise his jaws by felling trees, would soon fall behind in the struggle for existence. A person who has a weak organ—eyes, for example—suffers no serious inconvenience by undergoing such tests; for he can earn chronops in physical as well as mental avocations. It will now be seen that there are ample inducements for manufacturing commodities and bringing them on the market, so that those who purchase them need never be stinted. Even such occupations as require little or no expenditure of energy, thus bringing no functions into play, may be performed, for there are periods during which people must rest, as well as those during which they must exercise their structures and functions. The chronops are very flexible; for they may be earned for the direct performance of special duties, as wages are now paid, or the same results may be obtained by measuring the development of our structures, or both methods may be more or less combined. If a person performs a given piece of work, we know that he has exercised certain structures; and if certain structures have gained in strength, we know that certain work has been performed—so that it makes no difference in the end which method of measurement is adopted.   We now clearly see that agriculture, manufacture, and trade are forced to produce men, and not men to produce them, as in the economic category—men who by their virtue and morality will shame the angels and the Gods—and the conversion of the soil of our isles, endeared to us by so many sacred memories, into a hundred millions of human souls is the most gigantic manufacturing enterprise that has ever been conceived; and we must have manufacture before trade can flourish.

In this work I must avoid details as much as possible, not only for the want of space, but also to prevent confusion with the first principles of the scientific State.   However, I shall here give one illustration, which will be generally applicable to all enterprises so far as the principles are concerned.   I shall select a form of enjoyment through which our functions can be most fully and completely exercised, namely, aquatic sports.   The first principle of development being to know how to breathe and to have an abundant supply of fresh air, the watering-place must always be the most scientific resort. The man who makes the best use of his nose has the first chance in the struggle for existence, providing he intelligently carries out this invaluable piece of knowledge; but he who wears clothing has hardly any use for a nose—and so long as he attempts to remain a cooking animal, his nose is of little use for smelling purposes.   The necessary preparation for aquatic sports is to enjoy air and rain-baths in country districts, so that when I select aquatic sports I presume the sportsmen to be much more highly developed than economic animals.   Those who make the best use of their noses have the strongest lungs, the greatest vitality, and the most active skin, and without these essentials motive and mental development dwindles into the limbo of abstraction.   But bathing and boating do not embrace all the duties of the aquatic sportsman.   The boat-factories are not erected in localities where boats can be made most cheaply, or with the expenditure of the least energy, but where they can be

most profitably utilised for the development of the sportsman's structures. It would be impossible for him to devote
his whole time and attention to rowing and swimming, because, in his examinations, he would fall below the standard
in general proficiency, although he may obtain first-class
honours as an aquatic sportsman. This drawback may be
partially remedied by two methods: (1) the erection of a
boat-factory, and (2) by the establishment of a library which
contains the best works on aquatic sports. This list, however, may be indefinitely increased, such as by the erection of
gymnasiums, factories for the manufacture of other commodities, &c. Now, when the sportsman also engages in the
manufacture of boats and in the study of aquatic mechanics, ·
he exercises a larger number of his functions, and therefore also earns more chronops. If he wishes to make a
speciality of aquatic science, he will be an expert boat-
maker, as well as oarsman and swimmer; but unless he becomes a distinguished inventor, or accomplishes something
unusual in the development of aquatic sports, he will find
himself obliged to move into some other centre of industry
for the purpose of earning chronops in other branches, thus
also exercising other structures and functions. Following
out the law that the most moral men have the choice of
offices, the best sportsman, having attained the highest aquatic
morality, will be entitled to the choice of boats, while he is
using one, and other privileges pertaining to aquatic codes;
but here a counteracting law comes into force. If he attains
his high degree of perfection through the neglect of other
structures and functions, and can thus earn very little more
in his special department, he will be forced into other avenues
of industry. Meanwhile, also, some distinguished biologist
may want the land which he occupies, consequently also the
house erected thereon, for the purpose of conducting agricultural or biological experiments. It thus appears that an individual may have specialities or may not according to his
ability for earning chronops—that is, he may become a scientist or a philosopher, or both combined. The question now
arises, How is the oarsman to pay for the use of the boat?
The boat is a draft on a certain land area, and he is actually

utilising a piece of land while engaged in the exercise of such structures as are brought into play during the operation of rowing, just on the same principle as he utilises land for his food and clothing, and the same method of measurement must be adopted. Before discussing this question, the periods of development must be considered. In the utilisation of all organic materials—food, timber, clothing—the measurement, in the first stage, is by the actual weight of the material ; the second stage, the weight of the dry solids ; the third stage, the weight of the ashes ; the fourth stage, the weight of the phosphorus. During this process of evolution the approach to the perfect standard is gradual. Let us suppose in this instance that the quantity of land in the boat timber is measured by the ashes—and for this purpose, it is not necessary to burn the boat in order to ascertain its ash constituents. The species and weight of the wood being known, and the character of the soil on which it grew, the average quantity of ashes which the given species of wood will make is the standard of measurement. Let us now suppose that it has been ascertained by many years of practical experience that the average life of a boat is six years of actual use—that is, when all the hours during which it has been employed are counted, the sum is equal to six years. If now a certain oarsman uses the boat for such a number of hours as equals six months, he must be debited with one-twelfth of the area of land which has produced the boat. It may be said that the boat also contains iron or steel, perhaps also gold and silver, in its construction. Very true ; but this alters no principle, and does not complicate the calculations in the least. These metals are also drawn from land, and although they cannot be reduced to the ash or phosphorus standard of measurement, yet all that is required is that they should bear some fixed relation to the organic materials, which relation must be in proportion to the estimated abundance or scarcity of the respective metals. This standard eventually overcomes all the objections to the existing standards.

The above illustration is typical of all industries and tenures of office, and from it many important deductions may be drawn. The inducement to accept an office is that the officer may be able to earn more chronops. The most striking features of the situation are: (1) that no individual can become the owner of any commodity whatever, and (2) that the employments for women must be substantially the same as those for men. There can be no inducement for the individual to claim ownership in any commodity so long as he has access to its use. Every commodity is held by society in trust for posterity; and this principle is not altered by the fact that the individual may enjoy the exclusive use of any commodity. A farmer, for example, when he purchases a plough will enjoy its exclusive use, and will be charged with the total area of land required for its production. So also when an individual purchases a coat or a loaf of bread. The principle underlying these cases is, that society finds it to be to the advantage of all and to posterity that the individual may enjoy the exclusive use so long as no other individuals suffer thereby, or are concerned in claiming a share in the utility of the commodity. In short, all commodities must be perishable so far as posterity is concerned; for it is a crime to store up commodities for posterity, because the manufacture of such commodities is the most natural method for the development of the structures and functions of coming generations. Let them but have equal opportunities for access to the means of subsistence, and they will not want us to hoard up property for them; their only demand is that we exercise our functions, thus perpetuating increased health, strength, and wisdom. When I say that all property is robbery, I do not include provisions, or such commodities as are necessary to supply our temporary necessities. Another interesting feature in the situation is, that there is no dualism of Socialism and Individualism. This dualism is only an abstract idea, and has no existence in reality. The economic Individualist may be defined as the man who wrings a monopoly from the State, and then calls upon it to defend the property which he has acquired by plunder. Destroy Individualism and the economic State loses all its functions, and the structures must therefore also perish. No conception of Socialism could have ever existed except as a contrast to Individualism, and therefore the Socialistico-Individualistic dualism is a mere abstraction, having no existence in natural law. The economic State must thus also be a fiction of the same character. With reference to the administration of justice, economicism has not lived and died in vain if it has proved how we can enslave ourselves in the chains of Fashionism. In the scientific State the most moral men will be the rulers—that is, the men who make the best use of the land which they draw on for their support—so that scientific fashionism, or social restraint, will impel all subjects to morality, or to imitate the fashions of their rulers. The economic ruler is the person who is permitted to draw on the largest area of land for his support, so that the social restraint of the economic world impels us to immorality.

Industries which would soon be abolished, and others which would be developed, might here be enumerated, but these deductions can be so easily made that it is not necessary to go into details, the guiding principle being that each structure should perform its natural function. I may be permitted, however, to mention that, amongst all the institutions in the scientific State, the theatre is one of the most moral, for in it the performers can exercise an unusually large number of their structures and functions; and a critical audience gains immense advantages. It is sometimes said that the theatre is an immoral institution on account of the paucity of clothing which the actors and actresses wear; but this, being an abstract idea, is the exact reverse of scientific truth, and if the theory has any value, it only applies to those who have made a wreckage of their virtue. Clothing is a hindrance to gracefulness of movement, and obstructs what should be the most perfect and attractive structures in the world. It is no wonder that people are ashamed to display undeveloped structures which are unable to perform their functions. In the scientific State, no objection can be raised against engaging in a number of employments, because, all tricks and secrets of manufacture and trade being abolished, there is nothing to be learned in one industry which is not a development of principles found in another. The reason why the economic man cannot engage in many occupations is because there are so many tricks and frauds to be learnt.

The development of education in the scientific State should not be passed without notice. The basis of education, after acquiring the ordinary rudiments in the primary schools, is to know how to attach ourselves to our mother-earth, from which we derive our being and subsistence. Land monopoly being abolished, people will settle compactly in favourable localities. In a central portion of every locality of convenient size there will be a primary school with experimental grounds attached, also a gymnasium and other conveniences for developing the muscular, mental, and vital structures of the children. There will also be a library, museum, &c., for the use of the adult members of the school district. There will be no advantage in living near the school or other place of public resort, for the individual who has to walk, say, two or three miles is developing his structures and earning as many chronops as he who lives near and spends the same time in the library or the gymnasium. Even when the distance is, say, twenty miles, and the individual utilises a bicycle, the same advantages accrue; and if he has to go hundreds of miles by rail, his resting hours may be utilised for the purpose. This arrangement destroys all value attached to land due to advantageous location. A theatre, public hall, factory, &c., may be centred in the same locality, if the conditions are favourable. The school-teachers should be connected with the experimental grounds. As nobody will be engaged more than an hour or two daily in one employment, every school will have at least three functionaries : (1) the educator, who teaches the youths how to think; (2) the mechanist, who devotes his attention to the development of the motive system; and (3) the biologist, who will have to do with vital functions. The earnings of the children, which are now usually called prizes and awards, will be paid in chronops— the same coin with which their professors are paid. In their teaching capacity these professors have no connection with the State; they can only act as State officials when their chronopic earnings entitle them to examine their own pupils or those of other professors. The examinations must be imperatively uniform throughout all the districts in the State, according to arrangements made by the federal authorities.

Every individual has a right to demand to be examined on any subject at such periods as may be arranged by the authorities. These remarks, however, only apply to the consequent term of our physio-ethic ratio. With reference to the antecedent term, namely, the area of land required to produce our chronops, a valuator for the State will be required in each local district, who, in his private capacity, may also deliver lectures in the school. His duties will be exclusively confined to the means of subsistence, and he must know the relative abundance or scarcity of all commodities, and of the raw materials required for their production. It will be his duty to locate factories and other establishments, to project railways, telegraphs, telephones, &c., and a special branch of his department will be to make accurate estimates of the yields of crops, to distribute commodities, and to gather statistics pertaining to his department. An important branch of biology is dietetics, and a knowledge of chemistry is indispensable to the dietetist, and physics become an important branch of the mechanical sciences. Naturally, the district or primary schools will be feeders for the higher institutions of learning, which will be conducted on the same principles. All education, mental and physical, should have a subjective basis—that is, before we commence to expend our efforts we should understand the incentive, and then, objectively, we should be able to measure the legitimate conclusions.

The regulation of inventions is an important feature in the scientific State, and the same law applies to literary and artistic productions. The value of every invention is measured by its ability to develop our mental, motive, and vital structures and functions. A machine for setting type, for example, would have considerable scientific value, for the type-setter cannot exercise many functions at the compositor's case, and he would therefore demand high chronopic wages to induce him to follow the trade, and the occupation is not so healthful as out-door employments. Under any circumstances, people could not be induced to work more than an hour or two per day at this trade. On the other hand, a machine for driving nails, for example, would have no chronopic value at all, for many important muscles are brought into play in driving nails with a hammer. When the inventor receives credit for his invention he has no more control over it, and it becomes public property. The same rule applies to all works of literature and art. If the author exhibits great inventive genius, or even if a work is brilliantly written, the chronopic value is correspondingly increased. Good judges in these particulars would make excellent editors, for they would select only such productions as are worthy of publication; but a production may have chronopic value without being published. I would be in favour of attaching little value to knowledge acquired by cramming. The exercise of function consists rather in the use which one makes of facts than in cramming them into one's cranium. True knowledge comes out of us; it doesn't go in. Cramming is the abstract (objective) basis of knowledge; eduction is the scientific (subjective) basis.

It is important here to understand the new meaning which I have attached to the word philosophy. In PART I., I classified it under the abstract method of thinking; here it embraces the harmonious adjustment of all the sciences. The philosopher plays an important part in the scientific State. To him we will be indebted for the just arrangement of chronopic values. Although he may have but a superficial knowledge of the various sciences, yet he must know how to arrange them and classify them according to their importance—to trace the effects of each science and each form of activity to their effects on the means of subsistence and our methods of procuring the same. This is rather the reverse order, however. He should commence with the means of subsistence, and from this basis evolve all the sciences and all human activities. Our sharpest and most incisive critics should be the philosophers.

It may be urged that I have committed a breach of natural law in permitting those individuals to live who have fallen behind in the struggles for existence—that the weaker amongst the lower animals perish through starvation before the natural lease of their life is expired. This is a most important question pertaining to our inquiry, and although an answer may be given by the mention of one word, yet there is no subject whose origin is so imperfectly understood. In a word, I might say that man is the only religious animal, but this statement is imperfect and misleading, because our conceptions of religion are just as false as those which we entertain with respect to wealth. In defining the temperaments possessed by all animals—mental, motive, vital—I omitted to name a system which is of great importance in the economy of nature. The systems may be named either (1) to convey an idea of the structure, or (2) to convey an idea of the function. The names I have chosen direct our attention to the function; the ideas of the structures would be conveyed by the words nervous, osseo-muscular, and intestinal systems, instead of mental, motive, and vital. It must not be forgotten that each system acts and reacts upon the others —they are all mutually dependent on each other—and in their analysis, it is important to classify them with reference to their bases. In the ultimate analysis, both the mental and the motive system may be characterised as objective, while the system of which I am about to speak is the basis of subjectivity, and in all animals is the source of virtue, which must be sharply distinguished from morality. On account of the extreme objectivity into which civilised man has fallen, no word has yet been discovered to convey the functional idea of the subjective system, which furnishes the initial incentive to all legitimate activity. It is a portion of the network of the nervous system, and the idea of the structure is conveyed by the name genitals or genital system. The functional conception may be expressed by " virtuel system." There is as much difference between scientific and abstract virtue as between scientific and abstract morality. Although the close interdependence of virtue and morality cannot be denied, yet these qualities belong to different categories. Now, virtue is a quality which belongs to all the lower animals, but the development of its characteristics in man is true religion. A feature of special importance with respect to this question is the close historic relation which it bears to matters pertaining to our material necessities. Agriculture has always been a non-respectable pursuit, and on the same principle it has always been disrespectful to speak about the basis of virtue. This law, which pertains alike to the material and the spiritual well-being of our race, has been a necessity in the evolution of mankind; for all the avenues leading to perdition must be crowded. As the means of subsistence must be destroyed for the purpose of creating ideal wealth, so pure and true religion must be destroyed by that of the impure and orthodox type. This law is a necessity of universal evolution; for science could never have originated had man not have become an educated animal—that is, had he always followed the instincts of his nature. The development of abstraction is a natural law in the history of mankind. The instincts and intelligence of the brute are limited to the rearing of the offspring until the young is able to obtain the means of its subsistence; man can recognise his offspring during the whole period of life, and the evolution of sympathy must keep pace with that of intelligence. The brute can organise for the temporary preservation of the individuals of the living generation; man can look into future generations, and sympathy must conform with this intelligence. This is religion—the evolution of brute virtue. Man does not lose his brute virtue by being a religious animal'; nay, he develops it. Religion is the evolution of family life, and man's affection for his family need not abate because he loves all humanity; nay, it becomes intensified. Even our compassion for the brute may grow with our affections for our families and our fellow-feeling for mankind. Bring me a man with a wrecked virtuel system, and I will show you a man without virtue—without religion. Without this basis all our real pleasures and heroic deeds vanish, and our race becomes extinct: strengthen and develop it and the opposite results will be inevitable. The impulses and instincts which preserve our race cannot be separated from those which generate it. And no amount of intelligence can be substituted for this subjective basis.

Following out this train of thought we arrive at the only method for the solution of the population problem. Our virtue has become a tender plant, and it can only be strengthened by the same law as that by which those plants which furnish us with the means of our subsistence are strengthened. Protection and too much cherishing have proved a failure. The question of virtue is important from the standpoint that it aptly illustrates what cringent slaves to abstract ideas we can become. Our virtuel degeneracy may be measured by our adherence to the shame idea, just as accurately as our vital degeneracy may be measured by the number of cooks, butchers, and grinding-mills. When we are permitted to speak about the origin of virtue and religion without employing whisperable and circuitous language, and with the same freedom as we now speak about our mental or motive temperaments, the population problem will be virtually solved. The wildest of all abstractions is the idea that our bodies are dotted with shame spots, and that we must act as if we had no natural instincts. This abstraction has driven us to the most shameful extremes of immorality. The stronger our virtue, the greater is our natural tendency to display our parts, and the more suggestive of posterity the part is, the greater is the desire for such display, if the organs are not deformed. If man has attained his high state of civilisation by virtue of his having discovered a fig-leaf to hide his shame, then the barbarians and the brutes must be miserably low creatures for having no shame to hide. We must elevate our virtue to the standard of that of the brute before we can become truly religious ; and the only way to begin is to abolish the quack nostrums of priests and medical councils. What dirty mess have our hands been in that they must be kid-gloved from the public gaze? Would not ordinary knit hide the shame just as well? What should be the height of our collar that it may hide the shame of our neck and our ears? What should be the area and density of our moustache that it may hide the shame of our lip? What colour should our goggles be to hide the shame of our eyes? Should the shame of our blush be hidden by silk or satin goods? Must we hide our perfections in order that sapless and deformed mortals may have a pretext for covering their shame and their deformities? If we look into the question a little more closely we can see the effects of another law. Why do we gourmandise too much food? Because it is scarce, and because we are educated to the belief that we live to eat. The reason why we breathe too little air is because this most valuable of all commodities is abundant. If it were customary to breathe nothing but bottled-up Italian air, costing a guinea a bottle, the man of guineas would breathe too much; and if our most expensive foods became as abundant as air he would merely eat to live, and could scarcely be induced to indulge even to this extent. I have seen the most delicious peaches so abundant that the owners scarcely regarded them as being fit for hogs, while bacon was so dear in the same locality that the people imagined they were starving unless they had an abundant supply. In the same district hare and squirrel were so abundant that they were hardly worth shooting. The dwellers bordering on the Niagara Falls do not rise early in the morning to gaze on the grandeur of the scenery, as do the tourists; nor does the sublime, inspiring roar of the falling waters disturb their slumbers. In the British Museum stands the original nude statuary of some of the Greek Gods and Goddesses, and artists of both sexes are constantly seen sketching these delightful forms of human beings in which none of the parts are wanting. Are these artists less virtuous than those who never saw a nude statue? On the very contrary, their virtue becomes strengthened just in proportion to the success of the sculptor in breathing life into his divine forms of art, whereas the modesty of the novice becomes shocked at the supposed degradation of the scenery. Even more, the virtue of both sexes of artists becomes strengthened by working together and discussing the various perfections and imperfections of all parts of these images. If these statues gradually grew into real life, the virtue of the artists would be correspondingly intensified. Now, if these artists thoroughly understood physiology, phrenology, and physiognomy, they could relate truth about the history of ancient Greece, even though they had never read a single page. When similar access can be had to the Apollos and the Venuses who now live and walk our streets, true art and science will begin to flourish, and not till then. Were the Greeks less virtuous than we because they enjoyed no medical councils to prosecute their sculptors for immortalising the most perfect forms of our race? In this respect, the stage presents the most virtuous tendency of our day. The human form should be the most perfect of all objects, combining the greatest harmony of structure and function, and access to it by men of talent and virtue should not be denied.   Why should we be ashamed to walk on the street, or sit in the parlour, in the same attire as we wear on the stage ?   People who insist that the ideal should suppress the real should be pronounced, by all lovers of science, art, religion, and social reform, to be creatures of low and vulgar minds.

The first step to be taken towards the founding of a science of population is to snuff out that prince of humbugs, the Medical Council, so that scientists may have a chance to talk over these matters without fear and trembling.   The theory that man can evolve into a shame animal is exploded, just as effectually as the theory that he can evolve into a carnivore, a cooking animal, an orthodox religionist, an economist, or an abstract thinker.   Even if we could go through such a process of evolution, where does the advantage come in ?   What we want is men and women whose virtue is so strong and their forms so perfect and attractive that they have no shame to be hidden.   All these questions are evidences of our degeneracy, and prove that the more abstract theories are developed the worse they become, the greater waxes mental and social discord, and the more shattered and rotten grows our civilisation. The Greeks, who made physical exercise the basis of their educational system, were far ahead of us, and where have there ever existed such ponderous minds ?   We have, through force of necessity, obtained a vague idea that the social problem has something to do with land, or the means of subsistence, and yet we have to go back to Moses in order to get " original " ideas about land tenures.   For the regulating of population, I have no firm basis on which a system may be founded, for economicism has barred the door against such an inquiry.   It is for the good of trade that man should be regarded as a shame animal ; and it is a wonder that the government has not pronounced our noses to be an organ of shame in order to create an insatiable demand for nose-bags, thus opening up a new avenue for manufacture and trade. That the protective idea is a pure abstraction is amply proved by the fact that a handsome face is our most attractive part, and yet there is no special law to protect our cheek from the bombardment of a kiss. The absence of a special science for the regulation of population does not, however, damage my system. The chronopic account of the individual may determine the age he should attain before marrying, and the prospective births could in this way be determined just as accurately as the prospective deaths by life insurance companies. By this method, only the most moral men would be permitted to have families, and the most immoral would be weeded out. I am convinced, however, that scientific methods could be devised by which all individuals would be permitted to marry as soon as they arrive at the age of maturity, a guarantee being given that the size of the family would be regulated by the chronopic earnings, the most moral men thus having the largest families. With reference to those who have no ambition to struggle for existence, it is neither necessary that they should marry nor be requested to exercise their structures and functions. But social restraint, or Fashionism, will be so strong an incentive that institutions for the support of idlers need not be dreamt of. Religion, the sentiment of humanity, will force us to be kind to our unfortunate fellowmen. The poor we shall always have with us.

Some people may draw the inference from the principles of Humanitism that the most moral men and women should have as many children as possible, so that men should have several wives, and women several husbands. This inference is not logical, and can only be made from the purely objective standpoint. It must not be forgotten that subjectivity is the basis of objectivity, and quality the basis of quantity. The family must stand between the individual and humanity. There is no escape from this conclusion. The devoted, loving wife, and the cheerful, loyal family are forms of subjectivity which cannot be eliminated. We must rise above the brutes, not sink below them. The religious sentiment also inspires in the children respect and affection for their parents. Theoretically, the most moral man will have a title to the most moral woman, should the ages and temperaments be suitable, but as the individual has supreme control in all such matters, he is neither bound to live in himself nor in posterity, should he think life not worth living. The tendency to equality being irresistible, there can ultimately be no social distinctions. When a person arrives at the marriagable age, his moral character being summed up in figures, these numbers may be advertised, and suitable pairs will then enjoy the greatest possible facility for making acquaintances and contracting marriages. The choice of suitable partners is the bulwark of Humanitism. There can exist no struggle for existence between the sexes, so that it makes little difference whether or not their chronopic earnings belong to the same category. Should it be desirable to have only one classification, there are ample schemes for attaining this end. Woman is the subjective element of mankind, and if man is not worthy of our devotions, woman is.

With reference to standards of measurement a word is necessary, although this is not a question of vital importance in our inquiry. A chronopic unit may be established by taking any original production—and in Humanitism every commodity is a production of science and art—and value it according to the relation which it bears to our physio-ethic ratio, say 1000 chronops, the one-thousandth part thereof therefore being a unit. All other commodities may be traced to this basis. Any form of human activity which bears no relation to the means of subsistence, or our moral ratios, can have no chronopic value. If it requires great mental effort and nice discernment to work out these relations, so much the better for our intellectual development. I am bound, however, by the logic of my inquiry to drive scientific principles to their logical conclusions; and this suggests an enlarged basis of measurements. Our metric system, being based on an imaginary line drawn over the earth's surface, is an economic idea. What we must have is a standard which will measure the progress of humanity, for man is now to be the measure of all things, and the trade idea must become a matter of minor importance. If I were not bound by the logic of my inquiry, I would have some diffidence in discussing this question, due to the universal prevalence of the abstract idea that the living, and the prospective living, must be plundered that the dead may be honoured, venerated, and propitiated.

We must either plunder the dead to appease the living, or plunder the living to appease the dead. There is no other door for escape. The scientist allies himself with posterity against our ancestors; the abstractionist allies himself with our ancestors against posterity. To come to the point at once, the basis of all measurements is the dust of the dead. More strictly speaking, the ultimate standard is the quantity of phosphorus in the average human body. The unit of length is the diameter of a sphere which contains these ashes or this phosphorus. The unit of time is measured by the number of units of length which an average person grows in a given period. This unit of length must be so arranged that it is decimally related to nineteen years (the cycle of the moon) or twenty-eight years (the solar cycle). Nineteen years make a good basis for the measurement of time, because a person approaches maturity at this age. From this unit of length, all measurements of weight, capacity, surface, and work may be derived, and these again are related to the progress of humanity.

The science of humanitistic industrialism, which corresponds to political economy in the abstract category, is quite simple and requires little comment. It is not exerted *against* the laws and forces of nature, but in lines parallel *with* them. Scientific wealth cannot be produced by man's working against nature; for this would imply one portion of nature working against the whole, which is absurd—just as much so as it is to expect the hands to prosper by working against the whole body. The welfare of the structure as a whole is bound up with the welfare of all the parts, due prominence being given to each part according to the relation which it bears to the whole and each of the other parts. Economic wealth can only flourish so long as man is a supernatural being, and is not therefore bound by the laws of nature. If he is partly natural and partly supernatural, then the scientist must know the ratio between his natural and supernatural powers, if a ratio can be formed from finite and infinite things or numbers. If some people are more supernatural than others, this also affects our inquiry, and no basis for experimental inquiry can be obtained until the theologian answers these questions.   The physio-ethic ratio is not composed of unlike quantities: it is a ratio between land and land.   In humanitistic industrialism, it is convenient to classify the materials and forces utilised in the production of wealth.   Wealth is *produced* when a commodity (any material or force) is changed or moved in such a manner that it becomes more conducive to the fulfilment of man's destiny; for example, when soil becomes food, fish becomes soil, or a tree becomes a boat.   By this classification wealth may be produced without man's agency, and the materials and forces at work may be classified as follows :—1. Personal force (the force exercised by the individual) exerted upon the commodity.   2. Corporate force (that exercised by the united action of two or more persons) exerted upon the commodity.   3. Social[1] force exerted upon the commodity.   4. Ceternatural[2] forces exerted upon the commodity.   This classification gives rise to various combinations; for instance, personal force exerted upon social commodities, ceternatural forces exerted upon personal commodities,[3] &c.   All forces are natural, but I have been obliged to coin the word ceternatural to express all other forces, too numerous to be mentioned, which are not included in personal, corporate, and social forces—such as the heat and light of the sun, and all the forces in nature apart from man.   Matter and force being inseparable, the actual condition is force and matter exerted upon force and matter; but sometimes the predominating idea is force, and sometimes matter.   This classification is important, from the fact that it is the mission of science to cause work to be performed by the ceternatural forces, except when the human forces can be more advantageously employed to enable us to fulfil our destiny without violation of natural laws.   Agriculture, for example, should be conducted almost exclusively by the ceternatural forces, even though we can exercise our structures and functions by the artificial operations of husbandry.

---

[1] In Humanitism, there is no distinction between social and political force; for the State is society in its organised capacity.

[2] Latin, *ceter*, the remaining.

[3] The individual is the theoretic owner of those commodities of which he enjoys the exclusive use.

We are now in a position to ascertain whether the British Isles are over-populated or not. This question could not be fully discussed in our chapter on the science of population, for we had to ascertain through later chapters whether or not man has evolved into a dress animal, and this is again linked with the area of land drawn on to supply us with spirituous liquors. With reference to clothing, liquor, and tobacco, the calculations can now be easily made. As already shown, 0.60 acre are required to support an adult on a vegetarian diet, and in the United Kingdom there are 2.60 acres for each inhabitant, reckoning the population in adult equivalents. This leaves a balance of 2 acres to be devoted to clothing and luxuries. Of these 2 acres, 1 is cultivated land, and it is estimated that about one-third of the remaining acre could be profitably brought under cultivation. Of good arable land there remains 1.33 acre for each inhabitant, over and above the area required for food, without considering that a large percentage of the 0.67 acre could be profitably utilised for parks, plantations, &c., and especially for experimental grounds devoted to the raising of hardy fruits, nuts, and other useful agricultural plants. In making the following calculations, it should be borne in mind that they are based on practical and scientific possibilities, not upon economic theories. Considering the difficulty in estimating the length of time it would take to change from one habit to another, very accurate calculations cannot be expected, but so long as liberal allowances are made, there can be no scope for criticism. In changing from an economic to a scientific diet, much depends upon the mental condition of the subject. If he is a scientist, and is therefore thoroughly convinced that science is no humbug, the change may be made almost instantaneously; but to the economic man, who may entertain strong suspicions as to the efficacy of scientific principles, the time may vary between six months and a year, and if the physical conditions are also unfavourable, two or three years may be necessary. It will not do to abandon the consumption of meat without at the same time making a radical change in the remaining portion of the ration. Some economic vegetarians are of opinion that the evolution from meat-eating to the scientific or raw-food ration must be made through economic or cooked-food vegetarianism. This is a grave error, and the theory has been the source of much mischief. As for clothing, even taking economic man as he is, he would be much improved in health and strength if he went five or six months of the year without clothes, or at most with the barest apology for clothing, taking our English climate as the basis, although the same estimate would apply to colder countries. With one or two years' preparation, 80 or 90 per cent. of our population would become more vigorous by going three hundred days in the year with little or no covering, and all liability to catch colds would vanish, other health arrangements being properly attended to. I shall suppose, however, that physical exercise is made the basis of education, which means that the rising generation would wear nothing but gymnastic suits, if they wore clothing at all, and I shall give them two suits in the year—a light one, weighing $\frac{3}{4}$ lb., and a heavier one for winter wear weighing $1\frac{1}{4}$ lb., both being manufactured from wool. This estimate would also be a liberal average for the whole population, for many would not wear out one suit in a year. Taking the Oxford down sheep as the basic breed, a fleece will weigh 14 lbs. of washed wool, and it will require five large sheep to utilise as much land as the area drawn on to support the bullock, as discussed in our chapter on population, namely, 4.16 acres. Allowing 20 per cent. waste in the fleece, we get about 14 lbs. of wool per acre, or sufficient clothing to support seven men, so that 0.14 acre will be sufficient to clothe one man. Coming now to our drink bill, our statistics show that our beverages in the forms of beer, wine, and spirits are equivalent to 2 bushels of barley for each inhabitant per annum—that is, the equivalent of $4\frac{2}{3}$ gallons of proof spirits. This quantity of barley contains sufficient nutriment to make a sustaining food ration for about two months, the consumer being an adult. Our yield of barley is about 32 bushels per acre, so that the average Englishman imbibes one-sixteenth of an acre of land per annum. Considering that about one-third of our population are either total abstainers or very moderate drinkers, it will be seen that it does not take an inveterate "boozer" to imbibe as many acres of alcohol as the area required for clothing—or even for food. When a man drinks more land than he eats, and when it is considered that the area drunk is utilised for destroying his functions, instead of developing them, the sham of the abstract theories of land titles is made plain, and the necessity for a scientific basis becomes evident to all rational and unbiassed minds. So far as tobacco is concerned, I can find no estimates of the yields per acre in the morbid attempts made to grow it in England, but the averages for the United States, which I have before me in the agricultural report issued by the American government, will serve our purpose, for if we import tobacco, the transaction, like that of all others, is merely an exchange of land for land —or plot for plot. The annual average in the American tobacco-growing States for the past ten years is 738 lbs. of cured product per acre, and if we take the average consumption to be 5 or even 10 lbs. per head, the area of land drawn on is so small that it does not materially affect our calculations. I may also add that the average yield of lint cotton in the cotton-growing States, as given in the same report,[1] is 183 lbs. per acre, against our produce of 14 lbs. of wool from the same area, so that the acreage drawn on to supply us with cotton clothing becomes correspondingly reduced. Now, when we consider that the area of land required to support an adult in food, clothing, liquor, and tobacco cannot exceed 0.80 acre, there is no necessity in discussing the question of bed-clothing and minor articles; as for carpets, they are a positive injury to health, and are only used to display the extent of our robbery. Making the most liberal allowance, one acre of land will be the utmost quantity required for the total support of an adult, so that there still remains very nearly an acre per head of good arable land which nobody can want or use, and has therefore no value. Even after making liberal allowance for working horses, which, for a time, would be required for tilling the land and other purposes, and which would require little or no more land than that utilised to keep us in beverages and tobacco,

---

[1] Report of the Commissioner of Agriculture, 1888. Government Printing Office, Washington, D.C.

there still remains a surplus of land, so that by abolishing these luxuries, which have no scientific value, there would still be at least an acre of good arable land for each inhabitant over and above the area required to supply him with all the necessaries of life. With one or two exceptions, the United Kingdom is absolutely the most densely populated country in the world, and probably also relatively—that is, we have a greater population in proportion to the productiveness of the soil, so that, if other countries also lived scientifically, we could not export any agricultural produce.

I desire to lay special stress on the results of these calculations, because they illustrate an economic principle which has not yet been recognised by our abstract thinkers, and proves beyond the possibility of all doubt that economic wealth is based on scarcity. This principle is aptly illustrated by a discussion which I had a short time ago with a disciple of Henry George, the champion of the "single-tax" theory, or evolutionary Socialism. Said he : " When the land belongs to the landlords, and if its value increases £10,000,000, this increment goes into the pockets of the landlords; but if the land belonged to the State, the money would go into the pockets of the people, so that they would be richer by this amount." To this statement I made the following answer: "If the land belonged to the State, and if the increase in value amounted to the sum you mention, the people would be poorer than before, and their maximum wealth would be attained when the value of the land decreased so enormously that it had no value at all." Of course, he pitied me on account of my stupidity. The blunder which he made was in his attempt to enrich the people by producing for them £10,000,000 worth of scarcity, but he can surely be excused on account of his stupidity, for our illustrious political economists have all committed the same blunder. There can only be one conclusion, namely, that scientific wealth increases in exact proportion to the decrease of economic wealth. Under scientific wealth, our isles could double, or even treble, its population without depending upon foreign resources, and every inhabitant would be rich, happy, and prosperous, while under economic wealth we cannot support half of our present population, and

a large majority are poverty-stricken, miserable, and rebellious.
The dualism of science and economicism is now made plain—
the agricolo-economic dualism.   Agriculture cannot be encouraged except at the expense of trade.   If we grow all our own
articles of consumption, we cannot import them; that is certain.
"But," exclaims the economist, "it is an export trade which
we want, and the foreigner will then be compelled to send us
the value in money."   That is to say, all nations are going to
send us all their gold and silver, leaving none to settle their
home balances, and we are going to utilise all this bullion, of
which there are 500 cubic metres of gold and 18,000 cubic
metres of silver in the world.   These quantities would make
more trinkets than we would want to carry, and would practically destroy the value of these precious metals.   It requires no argument to see this fallacy.   If foreign nations
payed our exports in matches we would at once see the absurdity of the trade theory, but we are educated to believe
that gold is impregnated with some supernatural principle.
What a blessing it is that gold is scarce! otherwise we might
have no means of displaying our crimes of robbery.   Again,
we must have ironclads to defend our commerce, which means
that we must be eternally preparing for war, or fighting
battles, in order to defend monopoly, to furnish respectable
employment for the hounds of war, to strengthen tottering
governments, and to remove our surplus populations.   If we
must encourage trade, we are obliged to ruin agriculture and
remove the scientist; if we must encourage agriculture, we
must bankrupt monopoly and trade and remove the economist.   It is well known, moreover, that our customs and
excises are raised almost exclusively by tax on such commodities as science condemns—commodities which demoralise
the people and incite them to crime.   One portion of the
community perishes for want of adequate employment and
through licentious living, while another portion suffers the
same fate in their vain endeavours and keen struggles to force
the unfittest to survive.   A critical writer in one of our leading magazines [1] says: "Our peerages are frequently lapsing
for want of descendants, and it is not the wealthy rector, but

[1] *Westminster Review* for March 1889.

the poor curate, who is notorious for a numerous family." This simply means that semi-starvation is more conducive to virtue than is gluttony. Some imbeciles parade their "moral restraint" and urge the same virtue on the part of the poor in the restriction of their families; but it would be well to inquire whether this motive has its source in pride, impotence, or philanthropy.

In a previous chapter I stated that the removal of land monopoly led to the destruction of all monopolies, including monopoly of the conscience. We have seen how it leads to the abolition of monopoly in the products of the soil, and it is now in order to discuss the effects on the conscience. In order to get at the root of this question, it is necessary to understand the essential origin of priestcraft. Let us go back to a state of society in which there lived three persons with their families who obtained their livelihood by fishing. The names of these persons are A, B, and C. Says A to B: " Don't you see that I am a stronger man than you, and am able to drown you in that fish-pond ? " B answers : "I know it, but please don't." A : "Well, I'll tell you what I'll do ; you give me a good big fish every day, and I'll call you a free man." B naturally assents to this proposition, because he desires to live, and the arrangement suits A, because he would rather eat fish than strain himself by catching them. Now, B hardly ever gets to see A, and is led to believe that he is a very great and strong man, and is even growing stronger, especially after he comes to demand two fishes for his daily ration. B has now to struggle very hard to catch fish enough for himself and A, and when the wind blows and capsizes his canoe with its contents of fish, he logically arrives at the conclusion that he hears the voice of a greater man than A, and that this great man can be appeased by offerings of fish. C now steps upon the scene; he has a dream, and climbs up to the mountain-top early in the morning, where he confers with the great man whose breath upsets B's boat. Accordingly C goes to B and says : "I had a talk this morning with the great man whose voice upsets your boat, and this powerful and all-wise being told me that I was to get the fish which you sacrifice to him, and that I was to.

offer up the sacrifices myself; but the condition is that you must give me two fishes daily, one for myself and the other for the sacrifice, and if you hear the voice calling very angrily and upsets your boat every day, you will then owe me four fishes, two for myself and two for sacrifices." B, who is very desirous of appeasing the wrath of this great being who dwells in the skies, naturally assents to this proposal, and the contract is completed. We have now two functionaries, namely, A, the State; C, the priest: B representing the labouring classes. It is now evident that when every man is obliged to catch his own fish, none of these monopolies can have origin, and when all incentive to priestcraft is removed, the priest must go : the structure cannot hang together after the function is gone. To continue the illustration, which depicts actual realities: if A and C kept aloof from each other, they would both soon rust out, and B would then be monarch of all he surveyed; but they get into a wrangle about the division of the surplus fishes, namely, all which are left after B gets enough for his bare subsistence, by which means they keep one another from rusting, and their existence depends upon their success in keeping B convinced that they are supernatural animals, and are therefore able to drown him eternally in the fish-pond. When B ceases to believe this doctrine, he finds that he is a stronger man than A and C put together, and so orders them to catch their own fish; but they are not able to do so, not having learned the business, and having almost exhausted the fish pond. A battle ensues in which B becomes king, D becomes priest, and then history repeats itself. This is human history in a nutshell.

In the evolution of sociology, the economic State should also be attacked from the metaphysical standpoint. In the discussion of this phase, I accept Comte's "Law of the Three States," namely, that human knowledge first passes through the theological stage, in which volition emanates from an external source, then through the metaphysical stage, in which volition resides in the object, but exists independent of it; and, finally, through the positive or scientific stage, in which these abstractions disappear, and the human mind is satisfied with an investigation of the laws of phenomena.

Dr. Bridges forcibly illustrates this law of evolution in the following language: "Take the phenomenon of the sleep produced by opium. The Arabs are content to attribute it to the 'will of God.' Molière's medical student accounts for it by a *soporific principle* contained in the opium. The modern physiologist knows that he cannot account for it at all. He can simply observe, analyse, and experiment upon the phenomena attending the action of the drug, and classify it with other agents analogous in character." Let us apply this law to the economic State, the "soporific principle" being Royalty. The theological idea that political volition emanates from the skies has vanished from the minds of all rational men, but the State still exists by virtue of an inherent principle called Royalty, which serves the same purpose as the old theological dogma. There is this difference, however, that the theological idea, emanating from unity of volition, tended to greater harmony of thought; in the metaphysical stage, which now predominates, some people locate Royalty in the crown, others in the wearer of the crown, while others confuse it with ancient traditions and usages, these passing from monarch to monarch even when an ancient branch of the Royal family becomes extinct. We can thus see that the "soporific principle" of Royalty loses, through dispersion, the intensity of theological soporification; and the clinging to the institutions of the past is a theological phase of metaphysics. Royalty inspires loyalty. The sluggishness of the evolutionary process is exemplified by the change from monarchical to democratic institutions— such as in the United States of America. In the Aristocracy, where the rulers are removed from the masses, and where the dispensers of justice are robed in awe-inspiring fashion, the royal and loyal principle naturally abounds, there being an imitation of theological inspiration and reserve; but in the Democracy, where the humblest elector may booze with the highest officers of State, there still lingers the "soporific principle"—the confused idea that there is something supernatural somewhere in the State. And yet it does not seem that the mission of Social Democracy is to dispel the "soporific principle" in politics; for the Socialistic idea is that the State can accomplish anything or everything, despite the fact that, in the opinion of the Socialists, it is a purely human institution. That human institutions can perform supernatural functions is a dualism which the Socialists have not solved. It is necessary to discuss this phase of the economic State, for a scientific basis cannot be firmly established until the metaphysical figment is dissipated from the minds of men.

If we now take a critical glimpse of the whole situation, we shall find that the scientific and abstract methods are diametrically opposed in every particular. The abstractionist commences with the idea and sticks to it. It is therefore natural that the mind should be crammed with abstract theories, which form the basis of education, that education should be free, that mental employments should be most respected, and that all pursuits which pertain to physical occupations or to the means of subsistence should be despised. Now, the Humanitist also aims at mental power, but his method of attaining this object is the exact reverse, and may be classified in the following order: 1. Land. 2. Vitality (vegetal and animal). 3. Motivity. 4. Nervosity. In the abstract conception, this order being reversed, there is no basis for evolution; for nerve functions cannot be perfected without attention being paid to the preceding requisites in the order named. This abstraction is as old as theology, and could never have originated save through the illusion that nerve functions alone are immortal—that a function can be perpetuated after the structure disappears. The motto seems to be: "With all your saving, save the function; and as for the earthy and earthly structure, it makes little difference what becomes of that."

In the light of the scientific State, let us now discuss some of the weaknesses of the institutions into which we may more or less rapidly evolve. We should bear in mind that existing institutions are a muddled conglomeration of Individualism and Socialism. In our own case, we are Socialistic with reference to our Poor Laws, our Education Act, our Established Church, our Land Act, our Artisan's Dwelling Act, and our Librarian Act. The political question of the day is whether we shall develop the Socialistic or the Individualistic phase of our institutions, the strong tendency being in the line of Socialism.   The Individualist, as we have seen, is the man who hires A to work his plot, hires B, C, D, &c., to labour on other people's plots, and then, after piling up the products for himself and his "heirs and assigns for ever," establishes an institution called the State to protect his booty.   It requires no arguments to show that the further development of this phase must aggravate our social grievances, and must end in suicide.   But there is another phase of Individualism, or rather affected Individualism, known as the "single-tax" theory, propounded by Henry George and his followers.   This authority gives the land to the State—it should be said that he gives the State a monopoly of the land, for his ideal State does not hold it in trust for posterity—and then gives the individual unlimited scope to rob the State of that portion of the land called the products.   He assumes that the individual has unlimited right to labour, and engage others to labour, on all the plots which his strength can command.   In Georgic economics, the individual is permitted to take all the fish out of the pond, leaving the pond itself for the State ; or, the individual gets the milk and the State gets the cow.   Moreover, the fish-pond, or the cow, must, at the same time, constantly increase in value, for it is upon this source that the State depends to meet the ever-increasing wants of the growing population.   Such a violent dualism will solve itself without consulting Georgics.   Again, the object of Henry George is to open up the land for the people—to give them free access to the land, so that they may be able to employ themselves in agriculture should they not be satisfied with the current rate of wages.   In this manner, he aims at raising wages and thus benefiting the labourers.   These objects are precisely what his scheme will fail to accomplish ; for (1) he does not make the land free, for the individual has to pay the full rent or tax to the State, instead of to the landlord, and if the State does not exact as much as the landlord, it acts unjustly towards the people ; (2) labourers cannot be benefited all round by high wages, for this implies that commodities must be scarce, and it requires a supernatural effort to comprehend how labourers

T

can be benefited by scarce commodities. A demand for labour can have no other meaning than a scarcity of commodities. If a fraction of the labouring classes can be benefited by high wages, this is inconsistent with the "solidarity" of the labourers. Henry George also exempts the products of labour from taxation, all the State revenues being derived from the land. In other words, he exempts all booty and plunder, and raises his taxes from the exhausted inheritance of mankind. Moreover, he attempts to drive the poor man, who is not satisfied with his wages, into agriculture, without having any knowledge of or experience in agricultural pursuits, and if he has capital enough to build a shanty and commence operations, his money is sunk, and when wages rise again in his trade, he will abandon his plot, and thus bring agriculture into greater disgrace. This economist apparently fails to see that there is keen competition in agriculture, as well as in the trades, and if a labourer is induced to abandon his trade and engage in agriculture, the inference is that the latter is the more profitable. Any appreciable development of Georgic "Individualism" can only lead to a speedy reaction. The weaknesses of revolutionary Socialism now remain to be seen. Amongst all labourers, both Socialists and Trades-Unionists, the motto is "The dignity of labour." The mental is subjected to the physical, the learned professions and other intellectual pursuits being merely things to be tolerated—may be useful, but not necessary. This theory leads to two absurdities. 1. It infers the right to labour, the individual being permitted to labour on other people's plots and appropriate the products, which is Individualism, not Socialism. 2. It makes the individual an instrument in the creation of manufacture and trade, instead of making these the instruments in the amelioration of mankind. If these economists regarded labour as an instrument for developing our structures, they would have the basis for the evolution of a sounder system of economics; but they still cling to the division of labour, which means the straining some structures and the atrophy of others, thus developing a puny and deformed manhood. They aim at bettering the condition of the labourers by raising wages, or making commodities scarce, which fallacy

I have already pointed out.   When commodities are plentiful wages must vanish.   Wages and the "dignity of labour" can only be enhanced by some method of making abundant commodities scarce, which is an absurdity.   The producers make them abundant, and the non-producers, who have power to draw on the products, make them scarce.   This dualism has led to an inflated gratification of human desires, and hence to the social condition which now exists.   The Socialistic method is, that all individuals must become producers, and that they must also be instrumental in making the products of their own labour scarce; in other words, they must both produce and consume large quantities of commodities.   It will not do to restrict the consumption, for this restraint would tend to destroy wealth, labour, and wages, all of which must be increased.   In a recent public debate, held in St. James's Hall, London, between Henry George and H. M. Hyndman, the latter gentleman, who is one of the ablest of the Socialistic leaders, stated that under the Socialistic regime labourers would not require to work more than two hours daily.   This statement undermines his whole system; because, in the first place, the scheme would restrict the production of wealth, individual and national, for, according to Socialistic economics, all wealth is produced by labour, so that the more labour the more wealth, and consequently also the more wages.   Even if this end were attained, the moral effects would be most disastrous, for if economic men were obliged to work only two hours daily, they could not be prevented from rushing into all manner of wickedness, even if every man, woman, and child in the State were appointed a special constable, or organised into an army for the special purpose of restraining crime and preserving law and order. If the Socialistic regime were long in operation men would degenerate mentally without improving physically.   Men can never improve physically under a minute division of labour; and when all intellectual incentive or competition is also removed, they will rapidly degenerate, eventually leaving the world in possession of the brutes.   It is the scheme of the Socialists to remove all competition, while Henry George is a believer in the efficacy of this incentive.   If man had

nothing to struggle for he would cease to exist; and all
Individualistic forms of struggling are those of a miserable
pack of robbers, cutting each other's throats for possession
of the proceeds of their plunder. By all means let us have
competition—fair, just, and honest competition—but let
us compete for morality and virtue, and not for vice and
spoils. In the Socialistic State, there being a multiplicity of
officials, all must live in a manner suitable to the dignity of
their office—that is, they must have power to draw on large
areas of land for their support; and here will be a dualism
between the dignity of officialism and the " dignity of labour."
If labour becomes the more dignified, nobody would want to
be an official, and a law must be passed to force certain
classes of the community to be rulers, and this would give
rise to the gravest forms of tyranny. On the other hand,
should it be decided that officialism is the more dignified,
everybody would want to be an official, and, in the absence
of State laws or regulations, the least cunning members of the
community would be obliged to labour; then we would have
the same conditions which now exist. The ultimate weakness in all the demagogues who affect to be the champions of
labour—and the same weakness is also found in the champions
of Individualism—is that their theories are profoundly objective. They merely aim at bettering the material condition of
the community, or some portion thereof, and even in this they
have proved themselves to be miserable failures. Their little
schemes of affected subjectivity have been as false as those of
their objectivity. When our true subjective nature is once
discovered and acted upon, no such thing as false objectivity
can exist. All attempts to better our material condition must
fail so long as we entertain false conceptions with respect to
our spiritual nature. A true and lasting solution of the social
problem can only be effected through what comes out of us,
not through what is crammed in. The kingdom of Heaven
is within us.

All these contradictions and weaknesses are evaded in
Humanitism, and the whole range of political economy is
brought into unity and harmony. In humanitistic industrialism there are many details to be discussed which are beyond

the scope of this work.   However, I shall here consider a few objections which may arise.   It may be asked how many people know those laws of nature which they are called on to obey.   This question may be answered by asking another. How many people know the 25,000 laws contained in our statute-books, for the disobedience of which we are all held responsible?   And yet these are a very insignificant fraction of the laws we are obliged to obey.   How many of our best lawyers and judges know the one-hundredth part of our laws?   Through disobedience of human laws, the most moral men are essentially those who suffer; but through disobedience of natural laws, only the most immoral can suffer and become extinct, which is quite consistent with the law of evolution.   Human laws being invariably adverse to natural laws, it must of necessity follow that the most disobedient with respect to the former are the most moral men. It is virtually impossible for us to perform moral actions without disobeying human laws or committing a breach of social etiquette, which is, in many instances, the most stringent of all our laws.   This is a necessary condition in the progress of universal evolution.   Natural laws can be understood by logical deductions from principles, so that the best reasoners are those who must and should survive.   Human laws, written or unwritten, have no logical connection, and the conjecturer is more apt to escape punishment than the reasoner.   The only logical conclusion which the scientific reasoner can arrive at is this, that if he sees a human law stringently enforced, disobedience thereto implies conformity with some law of nature.   Even for the most revolting crimes, the real perpetrators escape punishment.   Obedience to natural laws implies the extinction of all other laws, human and divine.   Another objection to our physio-ethic ratios may be raised with respect to the absolute morality of the man who draws on the smallest area of land for his support, other conditions remaining the same.   This question may be illustrated by the difference in productive capacity of apples and strawberries, and it is well to bear this is mind, for it will cover all other apparent difficulties that may arise.   The difficulty here is not in giving an answer, but in giving all the answers.

1. Apples being less productive here. than strawberries, it does not follow that the same rule holds good in all countries with which exchanges may be made. 2. If the man who eats berries achieves the same results as the man who eats apples, the inference is that the former are equally nutritious, or otherwise suited to the consumer's organs and their functions, and if identical results can be produced by the utilisation of smaller areas of land, the law of morality is more strictly obeyed from two standpoints: (1) that the berry-bushes must have greater vitality, or (2) they are better adapted for intensive cultivation. 3. If we do not know whether berries or apples are the better adapted to our organs and their functions, what then? Science is dumb on this point; but, in the first place, a continuation of the experiment would enlighten us on the subject. Again, we must have apples because they keep fresh all the year round, and we know that fresh fruit is better than dried. This leads to another question, namely, Can berries be kept fresh for mostly a whole year? That they could be kept fresh depends upon the character of the labour required, and how much a person can earn while engaged in that labour, which is a matter of chronopic arrangement. It will also depend upon the inducements offered by the State for inventing a process by which berries may be kept fresh, or the inducements offered for demonstrating that a proposed preserving process is not effective. 4. It is not necessary to separate apples from berries in calculating the area of land. For instance, the fruit-grower, if he thinks he can gain an advantage, may grow strawberries under raspberry-bushes, and raspberry-bushes under apple-trees, so that the quantity of fruit per acre will determine the results, not the quantity of each kind separately. If the grower obtains too large a quantity at the expense of quality, he can gain nothing. 5. Should any variety of fruit which is very excellent in quality,[1] but defective as a yielder, be regarded as worthy of being perpetuated, it may be given to the individual to be tested, under such conditions that the loss of land area will be compensated by the value of the

---

[1] It should here be borne in mind that there is a wide difference between scientific and economic ideas of quality.

test.   This is a just arrangement, for such tests are beneficial to the whole community.   All such matters will be regulated by custom.   No objection can be raised which can destroy the absolutely just and moral character of our physio-ethic ratios; and it may also be added, which will be discussed in a later chapter, that if man is naturally an apple-eating, rather than a strawberry-eating, animal, there is a basis for evolution, for the necessary conditions exist for this process.   Another objection may be raised, to the effect that in the scientific State there is scope for all manner of frauds.   He who has arrived at this conclusion is not a logical reasoner.   The leading blunder which he commits is the inference that the scientific, like the economic, State is a necessary evil, whereas the direct reverse is the case.   Even were every individual perfectly trustworthy, the necessity for the scientific State is not diminished.   It is necessary to know the productive capacity of the soil, measured by the yield of crops, in order to ascertain the extent of population which the country can sustain, and not specially for the purpose of ascertaining how much a person eats or drinks, although the one estimate is logically connected with the other.   By the slightest attention to this question, we shall see that estimates may be made in two ways, the one being a check upon the other.   1. The amount of produce which a given area yields may be estimated by reckoning the quantity of pounds, bushels, or tons measured when the crops are harvested. 2. If every housekeeper had a memorandum of all the eatable commodities consumed by the members of the household or family, the sum of all these memoranda must represent the produce consumed by all the individuals in the State.   There being no money, there will be here a double check: (1) the books in the distribution offices will show the sales made to each individual, and (2) the individual must keep an account of his receipts and disbursements.   In this manner, all commodities can be traced.   When people's desires are few, the keeping of such accounts will be a trifling business.   Another objection may here be raised, namely, that people do not go to the distribution stores for all the articles they consume, much of the products growing in their own gardens, or manufactured on their own premises. Let us suppose a case in which it may be considered possible to perpetrate the greatest fraud. Farmer A has a field of wheat—and it must here be borne in mind that every commodity belongs to society, and that the individual can own nothing, except theoretically, as previously pointed out—and he desires to defraud society of a portion of the yield. In his attempts to do so he has only two courses: (1) to under-estimate, and (2) to over-estimate the actual yield. For a clear conception of this question, we must here again suppose a theoretic basis, namely, that the wheat, which belongs to society, must be deposited in a store-house under the control of the valuator, and then purchased back by A, the producer; or, rather, that portion is purchased back which A requires for his support and that of his family. This plan, however, is not followed out in practice; for A, instead of actually delivering a portion of the wheat, consumes it from his own granary; but by the theoretic basis we can understand the question more clearly. Let us suppose that the actual yield of his field of wheat is 100 bushels, this being the estimate which the State requires for the purpose of as-certaining the population to be sustained. Now, if A wishes to perpetrate a fraud, he must send in a higher estimate, say 110 bushels, or a lower, say 90 bushels. In the first place, it is the duty of the valuator to know as closely as possible the productive capacity of all qualities of land, and the dietetist will know within a small fraction how much a person can consume, and as A's accounts will show the kinds of work he has been engaged in, this estimate is facilitated; for the severer the physical labour, the greater will be the rations consumed. The valuator's accounts, as well as A's, will show the quantities of commodities purchased, and the balance of the rations must therefore have been produced by A. Here are two checks on A's fraudulent intent. But I will now even suppose that these checks do not exist. First, suppose A's fraudulent estimate to be 110 bushels, what then? If the valuator demands this quantity, where is A to get his extra 10 bushels? All that he can say is that he has consumed it. Very good; then he has drawn on more land for his subsist-ence than if he had sent in an honest estimate, and his physio-ethic ratio would be against him. Again, let us suppose that his fraudulent estimate now is 90 bushels. In this case, his piece of land has produced a lower yield than that of his honest neighbours, and consequently he is not so well paid for his results. Let us suppose another case. Two farmers, B and C, conduct their operations on different principles. B tills his land very industriously, works hard to keep down the weeds, and raises, say, 30 bushels per acre. C cultivates or produces a hardy variety of grain which exterminates the weeds, no tillage operations being therefore required, for the grain grows just as weeds now grow, but his yield is only 20 bushels per acre. Some will here say that, by the logic of my own system of morality, B is the more moral man. This conclusion is not logical; for, in the first place, C may have exercised his functions more in the studio or in the gymnasium than B has done in the field, and will also have earned more by producing his hardy variety of wheat than B has saved by drawing on less land for his support. This is a matter of chronopic adjustment, and will be regulated to suit the requirements of the times. The man who has come to the conclusion that there is room for fraud in the scientific State has made another illogical deduction. He has drawn the inference that agriculture, in the ultimate analysis, is a manual occupation engaged in by men of low mental calibre, whereas the whole world finally becomes an experimental plot conducted by biologists, chemists, and geologists, and everything produced will be the result of some experiment. These are not the men who will bring science into disrepute by robbing society of a dinner or two yearly. It may be urged, however, that there is still scope for fraud in the consequent term of our physio-ethic ratios. An answer may here be given from two standpoints : (1) the State officers, or the examining body, can only be composed of the most moral men and women who are above suspicion, and (2) they are tied by their own precedents. The latter statement, however, is only operative to a limited extent. When people rush too headstrong into a few employments, the reason must be that the chronopic list is badly arranged, so that the State will be compelled to regulate it in such a manner that all the avenues of industry will be equitably adjusted. In this manner, the individual can arrange his future plans without awaiting the results of State action. There is another check by which the examinations may be practically taken out of the hands of the State. I refer to matters of appeal. Should a board of examiners fail to give satisfaction to an individual, he can appeal to unofficered judges on the following principle. A person becomes a State officer by virtue of his excellence in a group of branches. Now, it is evident that there will always be unofficered individuals who will have a higher standing in one or more departments of this group, and they will be eligible as judges of appeal within their respective departments. It does not therefore follow that appeal judges are officers of State. Viewed from any standpoint whatever, the scientific State is absolute security against all forms of fraudulent practices; but my system should not be held responsible for any forms of indiscretion which may creep into the minds of economic men during the transition from the economic to the scientific State. Another objection may be urged against my law of marriage. It should be borne in mind that civilisation is not yet prepared for a science of population. This must be the latest of all the social sciences; for there is room in the world for all the population which can arrive for many centuries to come. It is a natural law that economicism should tend to depopulate a country instead of populating it, as the champions of economic theories have been attempting to convince us. The next stage of evolution will be the restriction of our inordinate desires, not a restriction of population, *i.e.*, providing we are not destined to pass through a mock trial of Socialism. When we begin to see that true wealth is based on abundance, not on scarcity, the natural consequence will be a curbing of our immoral desires. When I assert that the moral character of the individual *may* decide the age at which he should marry, it may be said that the faculties of some people do not develop until middle age, say forty years, and that such people should marry at the age of twenty-five or thirty, which is an absurdity. This reasoning is illogical, and arises from a misapprehension of the first principles of Humanitism.

It should not be forgotten that structure is taken as the basis or measurement of function, so that the first object is to have perfectly developed structures and organs; when these are present, the functions will be a matter of necessity.   A person's structures can be accurately measured when he attains full physical development, and for the accomplishment of this object the burden of work and exercise will fall on the young, while the more aged will be chiefly busied in mental occupations.   At the outset of our career, there will be some force in the marriage objection, but this is not the fault of my system. With every entry which a person makes in his business, he registers a portion of his pedigree, and if we had lived scientifically and kept a register of our pedigrees for the past two or three hundred thousand years, we could now calculate with unfailing accuracy what we would develop into mentally after we attained maturity.   When the sciences of phrenology and human mechanics are fully developed, a person's chronopic value, aided by genealogy, can be summed up without physical or mental examinations, such as we have now to pass through; but it is now impossible to predict whether phrenology or genealogy will be the more potent factor in determining our future achievements after we attain maturity.   The perfection of mankind implies the abandonment of the function-test, the structure being an unfailing measure thereof.   Besides, it should not be forgotten that a person can generate his quota of family even after the age of forty; or, taking man's natural average age to be at least a hundred years, the same object could be accomplished after the age of sixty years.

I have now completed my outline of the State under the new civilisation, leaving the details to be filled in by the reader.   We have yet, however, to discuss how the Church is related to the State; and lest it should be considered that my system is too objective, all that is required to correct this erroneous impression is to glance at the heading of the following chapter.

CHAPTER V.

THE UNKNOWN GOD.

BEFORE discussing the question of religion, it is necessary
that no misconception can arise with refereuce to the classification of our functions. It has been my misfortune to have
a vocabulary restricted to our abstract conceptions of things,
so that our language has no words to express the actualities
in the conditions which I have been investigating; but by
strict adherence to the context and by drawing a sharp line
between the abstract and scientific categories, no confusion,
I hope, will arise. It will have been observed that I have
used the word ethics, as expressed in the phrase "physio-
ethic ratio," to embrace all the functions of nerve-tissue—
morality, virtue, and religion. Morality is mental objectivity; religion is virtuel subjectivity. But it is necessary
here to understand how these qualities differ from those of
the brute. The virtue of the latter is limited to the impulse
which enables it to generate and preserve its offspring and to
protect its species. These instincts are the subjective characteristics of its nature, and form the basis of its objective activities. In this sense, hunger, for example, may also be classed
as a subjective instinct; but this quality is strictly limited to
the individual, and the Humanitist can take no coguisance
of mere egoistic impulses as playing a part in the ethics of
humanity. The objective activities to which hunger leads
only preserve the individual, and of itself would not lead to
the generation or preservation of the species. The question
now arises, What virtue does man possess which is not found
in the lower animals ? Man does not perpetuate his species
by virtue of any impulse differing from that of the brute.
In fact, he is a mere brute in this respect, and even less, for

the latter does not abuse the instinct which leads it to beget and preserve its young. Man's virtue is therefore below that of the brute, although there are individual exceptions to this rule both in man and the lower animals : some men's virtue equals that of the brute, and the virtue of some brutes is as low and degrading as that of some men. Man compels some males amongst our domesticated animals to degrade their virtue. If the preservation of our species depended upon man's intelligence alone, there would be no struggling for existence, and our race would become extinct. So long as our virtue remains below, or only equals, that of the brute, we must struggle for existence in the same way. Here now comes the point which distinguishes the Humanitist from all other schools of sociology. He holds that man, unlike the brute, is capable of finding his way beyond the existing generation and of scrutinising the future. This sympathetic instinct is religion. If the brute is virtuous, man is therefore super-virtuous. But we cannot be super-virtuous without also being super-moral. The morality of the brute is limited to such activities as are necessary for the obtaining of the means of subsistence. If it be a carnivorous animal, it cannot, on the average, run much faster than the animal upon which it preys. Thus we see that sudden changes in the environments may tend to the extinction of the species. Man, on the other hand, can—although he does not—so adjust his activities that he is largely independent of future contingencies, and in this sense he is super-moral. To carry out the illustration, he can exercise his structures and their functions to such an extent that he can run twice as fast as the animal upon which he preys. In short, if his sympathies are with humanity, his objective enterprises may be of such a nature that they tend to the preservation of our race. Man, before his fall, developed the germ of religion ; and this is the foundation of the edifice of Humanitism. The existence of family and of patriarchal instincts has been a necessary condition in the chain of universal evolution. With respect to the moral, we have fallen below the lower animals ; but, having the super-virtuous, the next step is the development of the super-moral. These qualities, however, are inseparably connected, and all our activities must have their ground in the subjective elements of our nature. We thus also see the importance of faith and hope as a basis of our actions. Faith must precede our works.

We are now in a position to comprehend the violent nature of the dualism which has been playing havoc in the Christian Church. Christians worship the past, and repose their faith and hope in the future. They expect future rewards for the virtue of giving to the past. They fail to see that they rob the future of those very rewards, material and spiritual, which they give to the past, and also expect for themselves. They give away the cake and keep it, and then eat it and hand it down to their "heirs and assigns for ever." Like all other schools of abstraction, Christians have the right ideas in the wrong category. In order to bring the spiritual and material into harmony, I am bound by the logic of my inquiry to generalise the ideas of the future, instead of those of the past. The generalised ideas of the past are called God, more specifically known to Christians under the name of Jehovah, or Christ. In order to avoid confusion, I am obliged to coin a word which expresses the generalised ideas of posterity, and which corresponds to Jehovah in the abstract category. I select the word *Mellos*.[1] It is a gross fallacy to suppose that we cannot know the will of Mellos, who presides over the laws of nature, and I have already shown his influence in the production of economic wealth. Mellos is not the Unknown God whom we ignorantly worship, but the God whom we intelligently plunder. It should be borne in mind that we have made all the Gods in our own image; but he who stops short at this point is not a logical reasoner. It is just as true to say that the Gods made us in their image. When we make a God, he influences our actions, and we are made by our actions. If we are inspired to exercise a certain faculty, that faculty grows—is created and developed. It is important to retain the word creation, to signify landmarks or revolutions in the great epoch of universal evolution. Creation is a potent throb in the life of evolution. If now we are inspired by the spirit of Mellos,

[1] Greek, *Mellos*, the future (One).

we are created by him, and he controls the direction of our line of evolution. Mellos is not an arbitrary being created by man, but a logical necessity of our spiritual and material existence.

Before further developing this subject, it is necessary to point out the dualism of science and religion, and show its relation to Humanitism. On this point, I must confine my remarks to the relation between science and Theism (philosophy); for no scientist can accept such a ridiculous farce as the Plan of Salvation, or any other plan which has its source in priestcraft. The scientist confines his inquiry to the laws of nature, observing that the same antecedents always produce the same consequents under the same conditions, and he then arranges and classifies these laws. The Theist affirms that a law implies a lawgiver. The scientist answers, How do you know? This is the whole question in a nutshell. The Agnostic grants the existence of a First Cause, but asserts that he can know nothing about it, which puts an end to all further inquiry. The Theist strives to learn all he can about his lawgiver, and invokes science and philosophy to his aid. The orderly succession of phenomena is granted on all sides, but the scientist wants to know if this condition is the result of a preconceived plan—if it presents evidence of design. To the orthodox theologian, a law seems to necessitate a lawgiver, but to the Theist, this merely appears to be the more rational explanation of phenomena, and is therefore related to his belief. To the scientist, the existence of a lawgiver is not a logical necessity : the existence of laws implies the existence of harmony in their arrangement and succession. The Humanitist demands an answer to the following questions : Is the lawgiver merely the First Cause of the phenomena which we behold, or of all the phenomena of the Universe? Should we strive to learn the First Cause of the limited or the unlimited phenomena? Which of the Causes should we worship? Who caused the existence of the lawgiver? Which is the greater power, the lawgiver or the laws? Which is the more rational belief, that the lawgiver or the law is self-existent? Should we obey the lawgiver or the laws? Which is the more credible—(1) that the lawgiver produced the law, or (2) that the law produced the lawgiver?

The position of the Humanitist is different from that of both the scientist and the Theist. It may here be said that the Humanitist is a scientist. Very true; but there are three classes of scientists: (1) the man who acts for the purpose of making money, (2) the man who acts for the purpose of appearing wise, and (3) the man who acts for the purpose of improving his conduct—who acts for the purpose of improving his virtue and morality, not for the mere purpose of knowing what these qualities are, although this knowledge is a necessary outcome of his inquiry. This tripartism may be resolved into a dualism, namely, objective and subjective science. The Humanitist is the subjective scientist; the other classes are the objectivists. The former acts from the conscience; the latter from the intellect. The subjectivist evolves intelligence from his virtue (religion); the objectivist attempts to evolve virtue from his intelligence. With this explanation, no confusion, I think, can arise. The position of the Humanitist is this: he obeys the law, and in doing so he reasons that he cannot offend the lawgiver. He does not start with the theory that there is a " preconceived plan," but inquires whether nature has any plans, studies the process of evolution in the historic pages written by human hands and on the rocks, and then drives the knowledge thus acquired to its logical conclusions, both with reference to the past and to the future—the beginning and the end of natural phenomena. The more he learns about the laws of development, the more he can comprehend about First Causes and Last Effects. To him the latter are the basis, for it is more important to know the effect of an action than the cause. He worships the Last Effect instead of the First Cause, and in this way he gets a step ahead of the abstractionist, for the First Cause is produced by the Last Effect. Hence we again see the necessity of worshipping our posterity, instead of our ancestry, if the word worship is to be retained. Pertaining to himself personally, the Humanitist is an out-and-out Agnostic, but he soliloquises thus: "If I obey the laws I shall bring children into the world who will be stronger, healthier, and wiser than myself, and through generations of faith in this process the time will come when my posterity will know something about First

Causes and Last Effects; but if I disobey the laws, no faith or hope in the attainment of this end can be entertained." The present conclusions which science must lead to are that the laws of nature are immutable, and that if other planets are older and more highly developed than ours, they may contain more highly organised beings—Gods—but it is our duty to obey the laws of our development, not those of theirs. When we enter the kingdom of Heaven and become Gods, then we must obey the divine laws. If we attempt to obey the divine laws now, neglecting those by which we are controlled, we can never enter the New Jerusalem. If the Gods are mightier than we, they have attained their sovereignty through strict obedience to the laws of their being and development; and the higher the Deity in the scale of existence, the greater is his control over these laws. When we learn to see that the divinities are produced by the laws which they obey—that the law creates the lawgiver—then we make a leap from the abstract into the scientific category.

It is now obvious that the Humanitist takes his starting-point from the future, not from the past, and that the Melloic religion cannot yet have inception, except as an undeveloped instinct. Those who have studiously followed my arguments will have observed that I have given vast opportunities for measuring the progress of humanity, mentally, virtuelly, and physically. But we have as yet no basis for measuring such progress. There are many things, however, which we do know—for instance, Mellos does not want an ever-increasing class of paupers, criminals, and lunatics—and it makes no difference in this particular whether we generalise the ideas of the twentieth century or the hundred and twentieth. By constantly measuring the progress of humanity in the various ways I have pointed out, it is obvious that every generation can see farther into the future than its predecessor, and thus Mellos becomes an ever-increasing entity, and not an abstract theory. In short, he may be reduced to mathematical rules, so that when we ultimately obtain control over our own destinies we become Gods. By constantly fixing our minds upon the will of Mellos, who inspires us to activity, he becomes our creator; and if we now personify the ideas of any living generation, or those individuals in it who are inspired to great deeds, under the name of the son of Mellos, it is remarkable how close is the analogy between the Melloic and the Christian religion. In both instances there are three persons in the Godhead, and all the believers or saints have become the sons of Mellos (John i. 12, and Rom. viii. 14, 24-26). Sin is disobedience of the law; and the resurrection will be when the graves are opened to pour out their spoils to be evolved into other forms of organic life. The power of Mellos is absolute; the will of our race is measured by the dust of fallen empires, although, as we shall presently see, Paul uttered a grand truth when he commanded us to work out our own salvation. He can be justified for accentuating the absolutism of the higher powers. Interpreted in the light of Humanitism, I have no hesitation in accepting the Scriptures as the grandest and loftiest conceptions that have ever been penned. I make these reservations, however, that I judge them by their general tone, and that their superiority is evidence of human degeneracy during the past twenty centuries. They originated in the true subjective spirit of the religion of humanity; but the Christianity of to-day is essentially the child of priestcraft, which Christ so vigorously denounced. Never has there been an age in which the conditions were more favourable for the creation of the loftiest ethical sentiments. The extremes of objectivity, the result of false philosophies, had been reached, the times were ripe for subjective reaction, and the facilities for spreading knowledge greatly increased in subsequent centuries. The experience of ages was concentrated in Alexandria, thus facilitating the compilation of the noblest sentiments of all previous ages. It was quite natural that the writers of the New Testament should give undue prominence to the function in preaching the preciousness and immortality of the soul, for this was quite in harmony with the abstract methods of that day, the philosophers having been champions of the Idea, and the body was therefore looked upon as a degrading lump of earthly clay. For those who understand the human nature of priestcraft, the errors, contradictions, and immoralities of the Scriptures can be easily traced. Many of the contradictory passages can be explained by ingenious devices, and the interpolations of the priests are quite in sympathy with this policy ; for it was to their interests that the common people should be obliged to come to them to have the intricate way to salvation pointed out—for a consideration. Before the days of printing, it was quite easy for the scribes to twist the Scriptures into conformity with their ideas ; and as for miracles, they come as a matter of course when everybody believes in them, and when there are no Huxleys or Tyndalls to inquire into the validity of the miraculous. Christ wrote nothing, and the words attributed to Him must be received with the greatest caution. In this inquiry we are only concerned in what has been written, the authenticity of the writings being a matter of little or no consequence. What I want to show is, that the writers display, by their tone, the spirit of humanity—an earnestness for the welfare of our race. The widening of the narrow range of Patriarchism, the extension of salvation to the Gentile, is intensely humanitistic. The Jews (subjectivists), who dwelt in the spirit, and the Gentiles (objectivists), who dwelt in the flesh, were all to become partakers of the new salvation. In the ethics of Humanitism there is nothing more touching than the words of Jesus (Matt. xviii. 1–6), when He placed a little child in the midst of His disciples, and the reward offered for merely giving a cup of cold water to "one of these little ones" (Matt. x. 42), clearly indicating the humanitistic tendency to forget the things behind and reach forward to the things before. Our children may be nearer and dearer to us, but our children's children are nearer and dearer to Mellos, and the claims of God should prevail. The child cannot cultivate a regard for humanity unless it learns to reverence and obey its parents. This duty is the first practical lesson in religion, and it is the duty of the parents to teach their children the truth by way of parable—to feed them with milk until they are able to assimilate solid food. The Epistles of Paul are a most passionate appeal on behalf of humanity, and nothing explains the truths of Humanitism more clearly than the parables of our Lord. Even the Lord's Prayer applies as forcibly to Mellos as to Jehovah. The doctrine of Hades is also prefigured, for those who do not enter the kingdom of the Melloic heaven must suffer eternal death. I will here quote a few passages which are in perfect sympathy with humanitistic ethics :—

"Come unto Me,[1] all ye that labour and are heavy laden, and I will give you rest. . . . My yoke is easy, and My burden is light (Matt. xi. 28, 29). Every kingdom divided against itself is brought to desolation ; and every city or house divided against itself shall not stand (xii. 25, 26). Every idle word that men shall speak, they shall give an account thereof in the day of judgment (xii. 36). Now the axe is laid unto the root of the tree : every tree therefore that bringeth not forth good fruit is hewn down and cast into the fire (iii. 10). I indeed baptize you with water. . . . He shall baptize you with the Holy Ghost and with fire (iii. 11). He cast out spirits with a word, and healed all that were sick (viii. 17). Neither do men put new wine into old bottles (ix. 17). What man is there of you, who, if his son shall ask him for a loaf, will give him a stone ; or if he shall ask for a fish, will give him a serpent ? (vii. 9). Thou hypocrite, cast out first the beam out of thine own eye ; and then thou shalt see clearly to cast the mote out of thy brother's eye (viii. 5). Do men gather grapes of thorns, or figs of thistles ? (16). The harvest is truly plenteous, but the labourers are few (ix. 37). There is nothing covered that shall not be revealed (x. 26). Whosoever shall do the will of My Father which is in heaven, he is My brother, and sister, and mother (xii. 50). It is hard for a rich man to enter into the kingdom of heaven : . . . it is easier for a camel to go through a needle's eye (xix. 23). If thou wouldst enter into life, keep the commandments (xix. 17). And Jesus . . . cast out all them that sold and bought in the temple, and overthrew the tables of the money-changers, and the seats of them that sold doves (xxi. 12). The stone which the builders rejected, the same was made the head of the corner (xxi. 42). Take, eat, this is My body ; drink ye all of it, for this is My blood (xxvi. 26, 27). Except a man be born anew, he cannot

---

[1] In comparing the Scriptural with the humanitistic allegories, the first person, or person speaking, in the latter case may typify Christ, Paul, or any of the disciples, for (like the children of Mellos), all are animated by a unity of spirit.

see the kingdom of God (John iii. 3). Every one that drinketh of this water shall thirst again (iv. 13). He that reapeth receiveth wages, and gathereth fruit into life eternal ; that he that soweth and he that reapeth may rejoice together (iv. 36). Search the Scriptures (v. 39). Work not for the meat which perisheth, but for the meat which abideth unto eternal life (vi. 27). I am the bread of life (vi. 35). My teaching is not Mine, but His that sent Me (vii. 16). The devil . . . was a murderer from the beginning, and stood not in the truth, because there is no truth in him (viii. 44). Ye are gods[1] (x. 34). I am the resurrection and the life (xi. 25). Except a grain of wheat fall into the earth and die, it abideth by itself alone; but if it die, it beareth much fruit (xii. 24).[2] If I . . . have washed your feet, ye also ought to wash one another's feet (xiii. 14). I am in the Father, and the Father in Me (xiv. 11). No longer do I call you servants (xv. 15). Beware of the scribes, which desire to walk in long robes, and to have salutations in the market places, and chief seats in the synagogues, and the chief places at feasts : they which devour widows' houses, and for a pretence make long prayers (Mark xii. 38–40). I am not ashamed of the gospel. The righteous shall live by faith (Rom. i. 16, 17). Gentiles which have no law do by nature the things of the law (ii. 14). There is none righteous, no, not one; there is none that understandeth, there is none that seeketh after God . . . (x. 10–18). Do we then make the law of none effect through faith? God forbid : nay, we establish the law (iii. 31). Where there is no law, neither can there be transgression (iv. 15). Being therefore justified by faith, let us have peace with God; . . . let us rejoice in the hope of the glory of God (v. 1–3). Shall we continue in sin that grace may abound? God forbid (vi. 1). The wages of sin is death (23). The law of the Spirit of life in Jesus Christ made me free from the law of sin and of death (viii. 1).[3] If we live after the flesh we

---

[1] A poor compliment to the gods of those days ; but the meaning is : Ye are the germs of gods.

[2] This passage, although not scientifically expressed, is the key to the humanitistic religion, and expresses the law of perpetuity.

[3] Those who obey the laws of nature do not feel that obedience is any burden, but rather a pleasure.

must die; but if by the Spirit ye mortify the deeds of the body, ye shall live (viii. 13). Behold then the goodness and severity of God (xi. 22).[1] So we, who are many, are one body in Christ (xii. 5). Let us prophesy according to the proportion of our faith (6).[2] Let every soul be in subjection to the higher powers: for there is no power but of God; and the powers that be are ordained of God (xiii. 1). Be diligent, not slothful (xii. 10, 11). We that are strong ought to bear the infirmities of the weak (xv. 1). They that are such serve not our Lord Christ, but their own belly, and by their smooth and fair speech they beguile the hearts of the innocent (xvi. 18). Render to all their dues (xiii. 7). God . . . shall bruise Satan under your feet (xvi. 8). God chose the weak things of the world, . . . and the base things, . . . and the things that are despised (Cor. i. 27, 28). The Spirit searcheth all things, yea, the deep things of God (ii. 10). Each shall receive his own reward according to his labour (iii. 8). If any man buildeth on the foundation of gold, silver, costly stones, wood, hay, stubble, each man's work shall be made manifest, . . . and the fire itself shall prove each man's work . . . (12, 13). All things are yours (22). Purge out the old leaven . . . (v. 7). He that ploweth ought to plow in hope, and he that thresheth, to thresh in hope of partaking (ix. 10). Knowledge puffeth up, but love edifieth (viii. 1).[3] If I speak with the tongues of men and of angels, but have not love, I am become a sounding brass, or a clanging cymbal; . . . love vaunteth not itself, and is not puffed up . . . (xiii. 1–4). Where the Spirit . . . is, there is liberty (2 Cor. iii. 19). For we know that if the earthly house of our tabernacle be dissolved, we have a building from God, a house not made with hands, eternal in the heavens (v. 1). He that soweth sparingly shall also reap sparingly; and he that soweth bountifully shall also reap also bountifully (ix. 6).

---

[1] Those who habitually obey the laws do not feel their severity.

[2] Faith strengthens our intellectual faculties, and the more intelligent we grow, the further can we penetrate the future, under the dispensation of Humanitism.

[3] That is, knowledge which has no subjective basis. The character of the objectivists is summed up in chap. iv. 19–21, and that of the subjectivists follows.

Though we walk in the flesh, we do not war, according to the flesh (x. 3). Before faith came we were . . . under the law shut up unto the faith which should afterwards be revealed (Gal. iii. 23). The law hath been our tutor to bring us unto Christ, that we might be justified by faith ; but now that faith is come, we are no longer under a tutor (24, 25). One thing I do, forgetting the things which are behind, and stretching forward to the things which are before, I press on towards the goal . . . (Phil. iii. 13, 14). Servants, be obedient unto them that according to the flesh are your masters . . . (Eph. vi. 5).[1] The labourer is worthy of his hire (1 Tim. v. 18). The love of money is the root of all evil (vi. 10). To the pure all things are pure : but to them that are defiled and unbelieving nothing is pure (Titus i. 15). Be ye doers of the word, and not hearers only (James i. 22). Whosoever shall keep the whole law, and yet stumble in one point, he is become guilty of all (ii. 10).[2] As the body apart from the spirit is dead, even so faith apart from works is dead (26). Whence come wars and whence come fightings among you? Come they not hence, even of your pleasures that war in your members? Ye lust, and have not : ye kill and covet, and cannot obtain ; ye fight and war; ye have not, because ye ask not. Ye ask, and receive not, because ye ask amiss, that ye may spend it in your pleasures " (iv. 1–4).

A deeper inquiry into the ethics of religions and philosophies would, I think, prove that the human mind cannot transcend the limits of experience ; and this question is easily solved when we transfer our thoughts from the abstract to the concrete category. At any rate, I have quoted passages enough to show that our conscience belongs to Mellos, not to Jahveh or Jove, and that the Humanitist is not come to destroy, but to fulfil. In a previous chapter, however, I stated that the Scriptures were very elastic, being susceptible of a great variety of interpretation. The orthodox theologian

---

[1] Under Humanitism there will be no distinction between master and servant, but so long as we are in the flesh obedience is necessary : the remedy is to get out of the flesh as quickly as possible.

[2] This passage is aptly illustrated by my remarks on vitality, namely, if natural laws are obeyed in every particular, except the use of our jaws, we shall perish just as surely as if we had disobeyed all the laws.

has now two doors for escape. 1. He may maintain that the Scriptures may be allegorically interpreted in such a manner that they harmonise with Humanitism ; in which case, however, they cannot be a divine revelation to man, for a writing which can be so interpreted as to defend both the thesis and the antithesis can neither be a revelation nor an inspiration. Nothing can be both true and false at the same time : man cannot come both from above and from below. 2. He may defend the negative side of the question, maintaining that such an interpretation is impossible, erroneous, illogical. In this case, he is obliged to disprove the truths of science, and show that Jahveh is engaged in solving the social problem, if there is such a thing to be solved.

Those of us who have had our eyes constantly fixed upon the past, and can only see the future as a hazy and formless void, must gradually remove the scales from our eyes by looking upon ourselves as the future from the standpoint of the past. Had William the Conqueror been a worshipper of Mellos, and had he been sufficiently sharp-sighted, compassionate, and powerful to witness and defend his subjects of the nineteenth century, would we now have weeping mothers and pitiful children imploring us for crumbs of bread? His being deaf to the voice of Mellos, and callous to the teachings of his gospel of peace and good-will, and of "liberty, equality, and fraternity," is no absolution for our atrocious crimes. On the contrary, though we may pardon his ignorance, the practical lessons of the past eight centuries make our guilt all the more unpardonable—our sins and crimes all the more flagitious—if we attempt to further develop and perpetuate his nefarious deeds. Could the souls of our ancestors rest peacefully in heaven if they were conscious of the wailings of their children here below—curses provoked by ancestral vices and transgressions?

We are now in a position to comprehend Humanitism in its three aspects — past, present, and future. In the abstract conception the starting-point is in the past, and the motion towards the present. A narrow idea prevails that humanity embraces merely the living in any given generation, the past and the future being matters of little or no human concern; that when we reverence and worship our ancestors we do our duty to the past, and when we merely provide for our own children, we do our duty to the future. The Humanitist takes his starting-point from the future, and gleans from the present and past all that is great and good and grand which will be pleasing in the sight of Mellos. All our thoughts and feelings are therefore Mello-ethic; all our doings Mello-physic. This is the only method through which unity, order, and harmony can be attained—an acknowledgment of the sovereignty of Mellos—and the only method through which we can fulfil our destiny; not as a matter of choice, but of necessity. We must learn to reverence an object not owing to its great antiquity, but on account of what it will become in the future, and the farther we can see into the future the more intense grows our veneration. But we cannot attain this end so long as we remain in a state of degeneracy. We may be pardoned for worshipping our great ancestors so long as they remain greater than we; but when we begin to draw our inspiration from the future, that moment we become free men, and our march towards salvation is secure. Let us worship the Last Effect which creates the First Cause, and the most bigoted Christian or the most ardent Theist cannot refrain from applauding our conduct. Mellos will inspire us with the noblest and most heroic of all subjective impulses.

Orthodox Christianity having had its day, and as it must fall with economicism, let us now drive some of the leading new religions to their logical conclusions, and compare the results with the legitimate conclusions of Melloism. At the outset it should be borne in mind that Melloism and humanitistic industrialism must go hand in hand—the one would be an impossibility without the other—while economicism must go hand in hand with Theism, Comteanity, Christianity, or some other new religion in the abstract category. What I mean by Christianity, however, is some development, or retrogression, of the existing forms. It can only continue to exist by relapsing into Romanism and terrorism; an enforced restoration of unity and harmony in religious thought—priestcraft, authority. This can be of no ultimate avail, because it would lead to the reassertion of religious individualism, and then we would be in the same position as we are now. Rationalism, Secularism, and Agnosticism are to be applauded for their services in attempting to thwart this disaster. A vigorous wave of Socialism would also be instrumental in preventing Sacerdotalism from raising its impious head. Strong as my sympathies are with the Irish people, yet, unless the Irish difficulty leads to the subjugation of Sacerdotalism, these magnanimous people will lay the foundation of their own. ruin. Priestcraft is the same all-devouring monster in all countries and in all ages. If people will not think with their brain, they must be compelled to think with their belly. These are all destructive forces; and before I pass on to discuss those forces which are both destructive and constructive, *i.e.* Theism and Comteanity, I may be permitted to define what I mean by Sacerdotalism. The word embraces a principle which is not limited to ecclesiastical matters; there is a "soporific principle" in the Church, as well as in the State, quite apart from the divine nature of its origin. In the ultimate analysis, the question resolves itself into this: Are we to pass through a stage of metaphysical ecclesiasticism—a condition in which the volition resides in the object, not emanating from an external source? We have ceased to believe in the divine origin of the State, and yet its "soporific principle" still appears to us to be superhuman. Are we to entertain the same belief with respect to the Church? An answer to this question will be a solution of the whole problem. If the Church ceases to be divine, and yet contains a superhuman power inherent in its stole or its pulpit, the metaphysical page of Church history will be a mere transcript of the divine page. The Church and State stand or fall together on this ground. If we are to have a religious war, it is quite possible that the metaphysical idea will lead us to excesses just as appalling and bloodthirsty as the religious idea has ever done. This leads to an inquiry as to the cause of such a disaster. I draw no distinction between the priestcraft of conformist and nonconformist bodies: the principle remains the same, although there may be variations in the degree. Authority under the rule of Sacerdotalism is identified with individualism amongst all religious bodies who engage priests to do their thinking in the region of speculative ideas. If there is more terrorism in the sacerdotalist, it is because he has more to defend : he must defend his property and his profession, while the nonconformist may only have his profession to defend. Under all circumstances, the priest will defend those rights and privileges through which he obtains a respectable livelihood, and possibly also lays something by for his "heirs and assigns for ever," in order that they may keep up that degree of respectability which is due to the family and the ancestry. In short, the priest must defend his monopoly. If we now connect this idea with the manner in which other monopolists act, when their "rights" are in danger, we may safely predict the turn of events. The principle of Sacerdotalism is therefore not confined to the Church. The priests are selling more indulgences to-day than at any previous period in history. Our Parliaments are workshops for the manufacture and sale of indulgences, and every doctor, lawyer, and apothecary is engaged in the same business. Our Royal Commissions are nothing but grinders of indulgences. The principle underlying the whole affair is this : there is an ever-increasing army who are pushing their way into respectable employments, which consist in the manufacture and sale of abstract speculations— indulgences—and which force us out of touch with nature and the laws of our being and development. We live in a fictitious region, and rebel when we are invited to draw upon the soil thereof for the gratification of our physical wants. What now do the constructive theologies offer us? The Theist thinks it rational to believe that a law implies a lawgiver, and accordingly he worships the lawgiver and disobeys the law. To him the law is an earthly affair ; and he wants to live a spiritual life. Naturally enough, this is an easy belief for him, because, his flock moving with him, he neither loses his employment nor his respectability. He condemns Sacerdotalism by urging his people to think for themselves in religious matters, and yet he retains the very system which coerces all religions and philosophies into suicide. He removes himself and his adherents from junction with the means of subsistence ;

and as he invokes science to aid him in abstract speculations, Theism becomes the most suicidal of all forms of Sacerdotalism. The Theistic laymen are called upon to think for themselves, and also to pay for getting their thinking done for them. The Theistic Church can only lead to the more rigorous type of Sacerdotalism, if it acquires sufficient wealth and strength. Comteanity is as yet free from the error of organised priesthood. The adherents of this religion, as worshippers of humanity, are quite consistent in their Socialistic proclivities; but they are still operating in the abstract category, and draw even more inspiration from the past than do all other religious denominations put together. Their leaders are the greatest educators of the people in all the leading branches of learning, which tends to draw men of ability and of keenness for inquiry. The weakness of the "Religion of Humanity" is the number of dualisms with which it is beset. In the first place, it is not the logical outcome of the positive philosophy of its founder. It aims to democracise wealth and aristocracise political power, which is a gross inconsistency. It acknowledges feeling to be the basis of thought and action, and yet subjugates the subjective element—women. The outward respect which it claims for women is not consistent with her political subordination. If women are unsuited for the performance of public duties, so much the worse for politics, and the only rational method of solving the dualism would be to reform the public duties. A profession which is not fit for a woman should be a disgrace to all honourable men. Women being the subjective element of mankind, it would be more logical to assert that they alone should enjoy the franchise; and if feeling is the mainspring of action, it must logically follow that women, under the same conditions and opportunities, are intellectually greater than men. Why do the Comtists, therefore, not strive to make the conditions and opportunities identical? In criticising the "Religion of Humanity," Professor Huxley says he "would just as soon bow down and worship the generalised conception of a wilderness of apes." For my own part, I would ten thousand times rather bow my knee to a wilderness of apes than to a city of Ripper Jacks. Man can never truly worship degenerating humanity—he has been attempting to do so for thousands of years—but he may worship the "generalised conception" of what he may become under orderly and progressive conditions. The object of our worship must be greater and purer than ourselves. Of all the new religions, the "Religion of Humanity" is the least dangerous, and if its adherents could devise some method of bettering our material conditions, they might justly indulge in the prospects of success. They have the right starting-point in ethics ; let them now seize the same advantage in economics and agriculture. If Comte had removed his "soporific principle" from the State as effectually as he removed it from other objects, he would probably have been the father of Humanitism.

Let us now examine the logical conclusions of Melloism. Every clear reasoner will see that it evades all the errors and injurious tendencies of other religions. Above all, it is sure death to priestcraft. The lawgiver is now subject to the law. This is no abstract theory, but is the logical necessity of the law of evolution, which must be universally recognised by every unbiassed observer, as well as by every earnest inquirer. The law is the eternal principle, and the efficient cause of all phenomena : the lawgiver is the subordinate cause of subsequent phenomena. The plain inference is, that it is our supreme duty to obey the law, not to inquire into the will of the lawgiver. Granted that the phenomena of our world is the direct effect of a lawgiver, it would be a contradiction to say that we can obey the law and disobey the lawgiver. Obedience thus becomes our supreme duty, and subjectivity the basis of all our actions. Conscience, *i.e.*, virtue and religion, is evolved from ourselves, and does not emanate from an external source ; it is the source of our morality. The Humanitist cannot separate religion proper from those other sources of emotion which inspire him to fulfil his destiny —such as the family, music, art, poetry, allegory, &c., all of which must also have a scientific basis. We are labouring under the delusion that civilisation is developing musicians. We must have a soul for melody before we can develop an ear for music. When we hear man out of tune with himself and with the rest of the world, when the pitiful strains of

social discord cease to rasp our souls, we surely cannot so far
forget ourselves as to say that we have musical ears. Let us
first have social symphony, and then tune our sacred lyre.
The most melodious strains are those which most inspire us
to do our duty. The mere imitative art of the present day
is without aim, and thus loses all genuine charms. Every
masterpiece of art should in some way represent the artist's
conception of our destiny, or some situation which inspires us
to fulfil'it. The plastic art of the present day is either an
economic idea or a representation of paganism ; it is capable
of enormous development in the scientific category. When a
great individual arises amongst us, Mellos demands the exact
configuration, not the head merely, but the whole body,
divested of such inutility as fashionable apparel : this is of
material importance in measuring the progress of humanity.
An ideal statue representing the human form should com-
prise all the knowledge which the generation possesses in
mechanics, anatomy, physiology, phrenology, and physiognomy—not a human being of the living generation, but one
whom, by prophetic inspiration and by our knowledge of
human progress, we can idealise in some generation of the
future. This will be another method of measuring human
progress, and such statues will be objects to be studied, not
merely to be gazed at and admired. These are statues of
Mellos, and should clearly indicate the generation to which
they are dedicated. Posterity will thus be able to detect our
errors, and evade them in their own representations of ideal
art. These statues are not ideal in the abstract sense of the
word, for when man begins to control the laws of his development, thus being able to calculate which of his structures will
tend to become extinct and which of them will tend to develop, prophecy then becomes reduced to mathematical rules,
and we become persons in the Godhead. Science has already
pointed out some of our structures which are tending to become extinct, but so long as society remains in a state of
anarchy, no accurate calculations can be made with respect to
our future—except that civilisation will be extinguished unless
we leap out of the abstract category. These statues will show
to future generations the degree of accuracy with which we

have interpreted the will of Mellos. The Goddesses will be an important and special department of this branch of art. Contrasted with this method of progress, I cannot forbear mentioning the decline in our statuary art as indicated by the statue of Achilles erected in Hyde Park, inscribed to the Duke of Wellington and his "brave companions in arms," which is the greatest satire on our religion as well as our art. We worship Jehovah, and depend upon the Greek Gods and Goddesses for our religion. To the scientific art critic, this statue, inscribed by our "countrywomen," indicates the true source and bent of our religion. The failure of Zeus and the subordinate Greek divinities consisted in their having human structures which were supposed to perform superhuman functions. The merit in the statue of Mellos consists in the structures being a correct indication of the functions; for he can have no functions which do not emanate from and are consistent with his structures. Jahveh is less scientific than Zeus, for he possesses supernatural functions without structures. Mellos, like Jahveh, quickens our subjectivity; but, unlike him, demands objective activity from his worshippers. The objectivity demanded by Zeus was through outward compulsion rather than through inward inspiration or necessity. Jahveh exclaims: *Obey!* The order of Zeus is: *Command!* The motto of Mellos is: *Obey that you may be able to command.* The religion of Humanitism, in its widest sense, embraces all sources of subjectivity. Poetry and allegory—call them fable, fiction, or parable, if you will—will all have a scientific and subjective basis. In aim they will not differ from painting and sculpture, and will bear some relation to our destiny, like all other objective activities. They draw out the beauty, strength, and harmony of our language, and are designed to inspire us to noble and virtuous deeds. We certainly cannot withhold our reverence from the noble heroes and heroines of the past, and as they have been animated by the spirit of Mellos, we must honour and respect their memories on account of their divine kinship. It may be urged that Humanitism is a system of downright cruelty concerning the bodies of our departed friends, and especially those who are near and dear to us by the ties of blood. The fear of

death is the legitimate parent of this fallacy. The model
Humanitist, conscious of having fulfilled his duties, does not
fear death ; and, dying, contemplates the sublime thought of
resurrection into the new life. The true man does not really
die ; he merely, by virtue of his old age, falls peacefully and
quietly into his last sleep, without a sigh or a pang. The
graveyard is the result of the fear that the deceased may
have entertained doubts as to his being in perfect unison with
the shades of his ancestors. Take away the fear of death,
and we destroy all sacredness in our ashes : only our spirits
merit our veneration, and they live twice—in our works and
in posterity. We should desire to live that we may do more
good. The desecrators of Mellos cannot live in their works
for lack of faith to perform, and they go down to Hades
instead of entering the kingdom of Heaven. Their name is
oblivion. Humanitism is a constant transition from the
unknown to the known.

It will have been observed that in Humanitism there is
a close and inseparable relation between Church and State.
The king rules by divine right, and all the officers of State
and Church derive their appointments and authority from
the same source. Mellos is lord paramount, enjoying the
monopoly of all the land and all the products, including man and his conscience, and his vicegerents and other
subjects on earth are merely tenants for life. It is the
duty of the tenant, at the expiration of his lease, to hand
over the land to his landlord without damage or waste, and
return interest on the capital invested instead of hiding his
lord's money.

I have now, I hope, shown clearly, though briefly, that
there is naturally implanted in the human breast sufficient
incentive to move us to generous and strenuous activity.
Despite the ruinous character of our political and educational institutions, let us cling to the faith which is inseparable from our being, and it will inspire us to fight the good
fight in the coming struggle for that form of "liberty,
equality, and fraternity" which will endure to the end of
time. Let us not deceive ourselves in the belief that mere
preaching can empower us to cope with the economic forces

which are organised and arrayed against us. We must work diligently and patiently, and fight courageously, not in the spirit of inhumanity or revenge, but in the spirit of love and piety and truth—yes, Truth, that awful tyrant to the disobedient and the factious, but that kind and submissive master to his obedient and faithful worshippers.

CHAPTER VI.

IN previous chapters I assumed the existence of matter and force, which may be regarded as unwarrantable. The solution of this dualism is not essential to the constructive phase of our inquiry, but may aid in destroying a number of the abstract theories with which we have been cursed. Our duties and responsibilities are now plain, and it makes no difference whether the ultimate phenomenon is matter or force, or any combination of these agencies. My claim is that I have solved all dualisms which have been regarded as such; but if I have discovered fresh ones, this is another question. A sharp line must be drawn between dualisms which produce harmony—such as centripetal and centrifugal force, pleasure and pain, &c.—and which are not problems for solution, and those which produce discord, which I have been engaged in solving. I have spoken of law, an eternal entity which we are called on to obey, and the question may be asked whether this phenomenon is some external volition, or some property inherent in matter, and yet possessing distinct and unchangeable characteristics peculiar to itself.

The word dualism I have used in more than one sense. In its widest signification, it means any twofold division, including an argument resolved into two sides: when it is a three-sided issue, I have called it a tripartism. In a more limited sense, we have the philosophic, the metaphysical, and the theological dualism. In philosophy, it is opposed to monism (materialism), mind being admitted to be distinct from matter, and a spiritual opposes a mechanical force. In reality this is a tripartism. In metaphysics and theology, the dualism is brought out more purely and distinctly: in the

former, the human soul is arrayed against lower animal life;
and in the latter, his Satanic Majesty is marshalled against
the Christian Deity, which, however, in modern Christianity,
is a very mild form of dualism. In another sense, all dualisms
may be said to disappear when man is conceived as having no
existence—either before his inception or after his extinction.
The scientist has only to do with actualities as he finds them,
not with abstract conceptions, but so long as he maintains
that the mind cannot transcend the limits of experience, he
cannot disregard the conclusions of abstract theorists. A
sharp distinction should be drawn between an abstract and
a scientific theory. The latter is quite legitimate, for it is
the logical conclusion of a truth, or truths, already investigated, and should a dualism arise, the question leads to
further investigation. When the result of an experiment
leads to two theories, one of which is capable of experimental inquiry and the other not, the failure of the tested
theory is strong presumptive evidence of the truth of the
other. The value of Humanitism does not depend so much
upon the truths already investigated, but especially upon its
ability to categorise theories in such a manner that they are
capable of investigation, and that, when investigated, they
bear a fixed relation to our physio-ethic ratios. It makes no
provision for investigation merely to satisfy our curiosity, but
awakens our curiosity to acquire knowledge which has an
established relation to our destiny. The dualism of matter
and force loses its supreme significance for the reason that no
scientist has yet proposed to burn a professional brother on
the stake for refusal to accept the sovereignty of matter, or
force, as the case may be. On the other hand, it has philosophic importance from the standpoint of the tendency of the
human mind to struggle after unity, order, and harmony.
In one sense, the word philosophy is used to express the
theorisings based upon the results of scientific investigations
—*i.e.*, philosophy begins where science ends. We have thus
three distinct meanings attached to philosophy; but in the
above-mentioned sense confusion may be evaded by substitution of the phrase "scientific theory." It would be as
futile for the scientist to attempt to get along without a

theory as it would be for him to attempt to get along without a Deity who presides over human destiny.

Before discussing the question of matter and force, it is necessary to understand the relation between organic and inorganic matter, and mechanical and vital forces. A sharp line must be drawn between organised and unorganised matter. We have (1) minerals (salts) which are inorganic, and (2) those which have become a part of plant-tissue—organised minerals or salts which are indispensable to plant and animal life, and may therefore also be called vitalised minerals. Again, we have such manifestations as sound, heat, light, electricity, all of which are reducible to one phenomenon. Reducing these to mechanical force, electricity, we have a corresponding vital force misnamed "animal magnetism." It should be called organic or vital electricity, it being also found in plants, and thus distinguished from inorganic electricity, or the electricity of inorganic nature—mechanical force. Electricity should be studied from three standpoints: (1) that of the inorganic world, (2) that of vital tissue, and (3) that of dead organic matter—post-organic electricity. We know nothing about vitalised minerals, for we must reduce living tissue to ashes before we can analyse them, thus restoring them to the mineral kingdom; but we can ascertain many properties of organic matter as a whole, embracing both the organic and the vitalised inorganic substances. We have now a basis for experimentation which enables us to add fresh testimony to some of the problems discussed in previous chapters, and which will, at the same time, give rise to new theories with reference to matter and force.

Let me here recur to my previous remarks with reference to the "vital electricity" in raw foods, which has been experienced by those scientific vegetarians who have closely observed the effects during the transition from cooked to raw rations. Not only do they experience an increased vital force, but also a pleasant coolness accompanied by an itching inclination to cast off the clothing and engage in vigorous exercise, sun, air, and rain-baths being specially pleasant and invigorating. It is known that plants have a rudimentary nervous system, and exhibit many characteristics common to animals, notably breathing, circulation, sleeping, and various complex movements; and it may be fairly assumed that the heat in cooking has, to some extent at least, the effect of dissipating the fluid which causes the vito-electric current in the nerve-tissue of the plant.

It pains me exceedingly to be under the necessity of citing some experiments conducted by myself. I might have given numerous experiments of my own in previous chapters for the purpose of substantiating the truths of other statements, but I purposely refrained, partly for the reason that I preferred quoting more widely recognised authorities, and partly because I desired to show that Humanitism is not a one-man institution, but rather a synthesis of the leading truths in all departments of science. My apology in citing my own experiments here is that, so far as I am aware, no similar ones have been conducted; but I do not ask anybody to accept my conclusions or my theories until they are also substantiated by men who enjoy greater public confidence than myself. In taking by the hands persons of highly developed nervous temperaments, I have succeeded in exhausting my nervous energy, by an effusion of the vito-electric current into the nerves of my subjects, to such an extent that my nervous prostration lasted two or three days. In restoring myself to my natural condition, I found that fresh raw fruits were the best agency, that raw meal proved more effectual than bread, that animal foods, even milk, operated still more slowly, and that hot cooked vegetables were probably the most ineffectual of all the rations which I tested. Electricity from the battery and alcoholic stimulants appeared to have no other effect than to deaden or distribute the dull sensation caused by the nervous exhaustion.

If this question were more thoroughly investigated, I am convinced that further proof would be added to the scientific theory, that ordinary electricity produces results akin to those produced by inorganic drugs, the effects therefore being merely distributive and disturbative, and not curative, while vital electricity produces nutritive and curative effects. One of my subjects used to return to me to have a "sick headache" cured. I make these observations to show the importance of vital electricity in the discussion of many problems pertaining to this inquiry. I shall not here drive my theories to their logical conclusions; otherwise I might be accused of creating worse superstitions than I have extinguished in the previous pages. Give me an instrument for measuring the vito-electric current, and I will reduce our thoughts and passions to algebraic formulæ. Vital electricity is also related experimentally to our habits of wearing clothes. Clothing can never be substituted for the natural hair with which our bodies were formerly covered. Hair is a regulator of temperature, moisture, and vital electricity, although it is not absolutely essential to our existence. It is unquestionable that clothing, dead organic matter, is poisonous to the system. No forms of artificial bathing can be substituted for the natural sun, air, and rain-baths. In our natural condition, we would run out to enjoy the genial showers, instead of protecting ourselves from them; and so far as shame is concerned, our natural instincts would teach us that we ought to be ashamed to attempt to escape such a wholesome enjoyment. Naturally enough, a shower of rain is injurious when we are wrapped up in clothing; and as for cold, the very thought of which makes us shiver, there is nothing more conducive to health and strength, and yet we must fight to get rid of it, just as we do with all other good things. Any person who has not experienced the invigorating effects of a cold atmosphere, in taking exercise uncovered after a cold bath, has not begun to enjoy life. In principle, wearing clothes is like eating cooked foods: in the latter case we deprive our vital organs of their natural exercise and stimulus, while in the former, the same remarks apply to our skin. Cleanliness, however desirable, is no substitute for this violation of natural law. What is probably still more injurious, clothing breaks the natural electric continuity between our bodies and the external world. So long as we continue our meat-eating habits, it is folly to expect genuine reforms in this direction. Perfection in health consists in not being able to have a conception of the temperature of the atmosphere or other medium which surrounds our bodies without the aid of a thermometer. This truth may be illustrated by the fact that if we depended entirely upon our face to measure the temperature of the surrounding atmosphere, we would be far from the mark in our calculations. Two or three hours' daily exercise in the open air, bodies uncovered, would cure more diseases, other sanitary measures also being duly attended to, than all the doctors in Christendom. The air and rain contain vito-electric forces which cannot be found in the bath-tub. An important distinction exists between organised and mineralised waters. The greatest nonsense that has ever been uttered is the theory that we can, through milling and cooking, deprive our foods of their natural salts, making up the deficiency by drinking mineral waters or swallowing mineral drugs.

The theory that matter cannot be changed into force, or force into matter, must be broken down. The collision of two equi-sized planets would generate sufficient heat to dissipate them into their original mist. But we need not depend upon such far-fetched illustrations. When we see ice converted into steam and steam into ice, we say that matter is converted into force and force into matter. Ice is potential force, steam potential matter, and water potential force and matter. This, like electricity, is mechanical force; but what is spirit? Let us hear what Bible commentators have to say about the following passage, namely: "The Spirit of God moved upon the face of the waters" (Gen. i. 2). The following theories have been advanced:[1] 1. A violent wind. 2. Elementary fire. 3. Rays of the sun. 4. The angels. 5. A certain occult principle, the soul of the world. 6. Magnetic attraction. 7. The Holy Spirit of God under the notion of a wind. The leading thought in these ideas is matter in motion, No. 6 being a submission to scientific ideas. Was this Spirit of God *something* that moved or *nothing* that moved? An attempt has been made to solve this dualism by the theory that intelligence is not a function of matter, but a certain somethingness or nothingness which moves, or is unmoved and unmovable, outside of matter. If such a thing exists, and if it can be moved by our importunities, all scientific inquiry must cease; for when the scientist importunes it to extinguish abstract theories, and the abstractionist im-

[1] Clarke's Commentaries.

portunes it to extinguish science, a confusion would occur which must derange the natural order of phenomena. My position is to show that all actions which emanate from the abstract category have proved a failure in the attainment of certain ends, and that the only recourse is to try the scientific category as an experiment.

When we once see that all matter is potential force, and all force potential matter, the foundation of a monistic phenomenon is 'laid. This phenomenon in its passive, negative condition manifests itself to us as matter, and in its active, positive condition manifests itself as force. It is all things actually or potentially. It will not do to say that heat expands matter and converts it into force, for expansion, or the removal of pressure, may have been the original source of heat—or the reverse the original source of cold. The life-history of inorganic matter is that it evolves into the organic, then reverts into the inorganic state, and so on *ad infinitum*. This law is not altered by saying that certain forms of inorganic matter — *e.g.*, clay, gold, silver — never become organised, or that certain organic elements—*e.g.*, nitrogen— do not revert, for we have no reason to believe that they have never reverted, and never will revert, or that traces of them may not be found in all objects, although we may not be able to discover them by chemical analysis. Indeed, the tendency of law seems to be towards the organisation of all the mineral kingdom. The conclusion is, that the vito-electric current is a subtle fluid—and we even literally feel it stream out of our bodies—which is capable of extraordinary attenuation and voluminousness. It is amorphous and proto-plastic, contains all the elements of vital tissue, is more viscid at low, and more volatile at high, temperatures, is transparent and colourless, and permeates all atmospheric space. Varying with the temperature, it becomes more or less rapidly renewed, the dead residue falling to the ground to be reorganised or to revert to rock, and go through another life's cycle. In this manner, the same elements which compose the vito-electric current in one cycle may become rock, inorgano-electricity, or vital tissue in the next cycle, and thus matter is being constantly changed into force, and force into matter. This theory explains many obscure problems in science, and is not impracticable as a basis for investigation. It explains that all spiritual and intellectual phenomena are due to the vito-electric current. It cannot therefore be said that law is force or matter, but a transition—a process of evolution.

CHAPTER VII.

In every system of ethics and industrialism the following relations are of specific importance :—1. Free-will and necessity. 2. Egoism and altruism.[1] 3. Man and woman. 4. Master and servant. 5. Education and instinct. 6. Nationalism and internationalism. Each of these questions should be the subject of a special chapter; but (1) owing to the restricted scope of this work, and (2) owing to the close and logical relation which they all bear to the principles already laid down, they may here be briefly disposed of.

1. *Free-will and Necessity.*—Of all philosophical problems, this one is least understood, and the theories advanced in all ages have been sources of great discord and confusion. If we could measure universal law, and also that portion thereof which belongs to our race, man's will could be reduced to a mathematical ratio. If it be our will to do those things which law coerces us to do, then the dualism of free-will and necessity becomes a nonentity—a fiction. It cannot be said that we exercise our will when we do what we are obliged to do; and yet when it is our will and good pleasure thus to act, it may be said that we possess unlimited free-will. The measure of our will resides in our power to disobey the law. The confusion of our abstract speculators is caused by their extreme individuality. They will tell us, for example, that they are perfectly free to eat cooked food or wear clothing; but from the humanitistic standpoint this so-called freedom is pure

---

[1] Altruism, originally from the Latin, has been borrowed by Herbert Spencer from Comte, and means the antithesis of egoism. Egoism is downright selfishness; altruism, a tender regard for the welfare and happiness of others.

fiction—an absolute nonentity—for a repetition of such conduct, extending through many ages, would end in the extinction of our race. The freedom of the will is closely allied
with the law of evolution. This principle may be illustrated
by the reasons why man has failed to become a carnivorous
animal, despite the attempts of so many centuries. If he
had, during the past two or three hundred thousand years, so
exercised his structures by running that he became enabled to
capture his prey, and then tear it with his jaws, as do the
carnivorous animals, he would now probably be a carnivore,
providing those human beings who were not able to obtain
their livelihood in this manner were systematically weeded
out. The hunter, the butcher, and the cook have, however,
thwarted our will in the process of evolution towards carnivorism. As a basis for evolution, we must have the subjective desire accompanied by corresponding objective activities,
and our structures with their functions must gradually become
accommodated to the change of environments. Free-will can
only be attributed to the lower animals in as far as they will
to do that which they are compelled to do by virtue of the
law of their being and development, and here no dualism can
be engendered. If they will to disobey this law, they become
extinct; and there is nothing in man which, in this respect,
differentiates him from the so-called brutes. Man, however, as
a super-moral and a super-virtuous animal, can exercise greater
freedom of will than the "dumb brutes," because he can bring
the law of evolution under greater control. Strictly speaking,
there is no such a dualism as free-will and necessity; but by
making our pleasure and happiness consist in obedience to the
law, we may learn to act as if we enjoyed absolute freedom of
the will. This dualism, like that of Individualism and Socialism, must be categorised as a nonentity.

2. *Egoism and Altruism.*—If this can be construed into a
dualism, my whole plan of salvation must be pronounced a
failure. "Live for others" is a motto of the Positivists;
and the motto "Each for all, and all for each" is rapidly
gaining ground in the Socialistic world. Such phrases are
very popular, and they sound very charitable and merciful.
They appear to be very altruistic, but they will not stand

scientific analysis. Charity is universally regarded as a virtue ; and, as we saw in a previous chapter, some people must suffer injustice in order that charity may abound amongst others. How can we "live for others" when, under economic rule, we cannot live for ourselves? The whole gospel of economic altruism may be summed up thus : "What is your will that men should do unto you?" "Why, give me money, of course." "Well, go therefore also and likewise do even so unto them." This is the result of economic forces, and then we must hire preachers to advise us to "live for others." Here is where the dualism comes in, which will solve itself despite the efforts of all preachers and other abstract theorists. The fundamental error in economic altruism is, that we are obliged to rob posterity in order to give to the living and the dead. Is that altruism? An ethical rule should have universal application. In order to show the failure of economic altruism in this respect, let us take an illustration. Of three persons, A is rich, B is in moderate circumstances, and C is poor. Now, if A lives for B, and B for C, whom must C live for? Seeing that C cannot live for himself, he certainly cannot live for others. If A, B, and C are all moderately well-off, each thus being able to live for himself, there can be no altruism : therefore the poor must exist in order that the rich may exercise the virtues of altruism, which is downright egoism. It will not do to say that nobody can live for himself, that everybody is dependent upon others for a living. The standpoint is that everybody is equally able, or has equal opportunities, to live for himself, and any exchange of services cannot therefore be regarded as egoistic or altruistic. Let us now take an illustration from the spiritual world. The mother loves and nurtures her son, and we will therefore suppose her to be inspired with the true altruistic principle. But the son must also be inspired with the same spirit, and must therefore reverence and obey his mother. The mother derives self-gratification in loving her son, and if she also expects the reward of reverence and obedience, she cannot be altruistic, for altruism must be performed without expectation of reward. Therefore the son must practise egoism in order that the mother may indulge in the pleasures of altruism, which is altruism. Thus we see that all attempts to construct a dualism must end in failure. Let us now apply a little humanitistic logic to this question. In the first place, everybody who comes into the world does so by the consent of society and of humanity, and it therefore makes no difference what accidents befall him, whether injuries be self-inflicted or otherwise, he does not depend upon charity, and can receive nothing but justice. In the abstract category, it may be said that there is scope for altruism in cases of accidents; but the same rule implies egoism on the part of the misfortunate. If it is altruism to give, it must be egoism to receive. The Humanitist accepts, as the basis of his actions, the fundamental principle that the desire of Mellos is that the most moral men should enter the kingdom of Heaven; and when his individuality is thus lost in the will of Mellos, he may be provisionally regarded as being intensely altruistic; but one-half of the story yet remains untold. At the same time, he must receive from Mellos all that is required to enable him to execute the divine will in the most efficient manner, which is, provisionally, an egoistic impulse. He knows that it is only through obedience to the law that he can bequeath to his children those virtues which fit them for divine inheritance. The process is therefore a constant giving and receiving—an eternal efflux and influx—and the ego-altruistic dualism thus becomes a pure figment. The governing of ourselves from the standpoint of the past has driven us into the dark abyss of egoism; the governing of ourselves from the standpoint of the present would sink us into the lowest egoistic depths, which can only be evaded by government through representatives of the future. To the Humanitist, the word justice comprehends all that the abstractionist can embrace in the words egoism, altruism, charity, and mercy. An altruist cannot exist without an egoist.

3. *Man and Woman.* — The relation between the sexes, especially between man and wife, must ever hold an important place in all ethical and industrial systems. It is stated in a previous chapter that there is no essential difference between the employments of men and women. While it is true that the law of development in both sexes is identical,

yet the same end can be attained by different means. The nature of the differences in the occupations of men and women can only be intelligently understood by a consideration of the ethical distinctions of the sexes. Having seen that woman is the subjective element of our race, it naturally follows that she should not engage in such daringly objective enterprises as those pursued by our sex. Under Humanitism, all the professions and other intellectual pursuits will be admirably suited for all womanly instincts, and the physical occupations, in proportion to the decay of menial employments, will be gradually opened up. That the women of the present day cannot understand politics and economics is the highest compliment which can be paid to them, and they are entitled to the same compliment with respect to the professions. Put these occupations on a common-sense footing, and they will be admirably suited for women; and our sex will have fearful competitors. By virtue of our moral courage, our objective activities, our sex should be able to win the love and esteem of pure-hearted and noble-minded women. But whence are we to draw our deepest inspiration—our most subjective incentives? From the fair sex, of course. Upon this question there is also too much individuality in the minds of our speculative thinkers; they are in the habit of regarding the individual to be the atom of a link separated from the endless chain of humanity. Man and woman, notably husband and wife, are intrinsically one person, the latter being the subjective and the former the objective element of one nature, and as subjectivity forms the basis of life and action, the part which woman plays may be readily conceived. There is another reason why men should regard women as the basis of their thoughts and actions. In matters pertaining to women's virtue, it is well known how prone men are to act as the aggressor, and so long as our sex assumes an air of superiority, they can act aggressively without shame or remorse; whereas, by cultivating a feeling of reverence and obedience, we are checked in our wicked career. The great man always worships the subjective. He who casts a supercilious glance at a woman is bereft of all virtue, all modesty, all shame. Woman is the long-sought philosopher's stone—the Jevonian instrument, sought by the Utilitarians, which measures the vibrations of our pleasures and pains. In the dim vista of the future, I already see the statue of Queen Mellos gazing meekly upon her king as he stoops to kiss her faithful hand; and my only bequest is that my dust beget the flowers that wreathe her sacred brow. Man and wife do not stand to each other in relation of inferiority and superiority; for objective activities should be an inspiration to greater subjectivity, just as much so as the reverse of this order. Where mutual relations exist, there can be no dualisms. I have no sympathy with those who counsel the subordination of women—those who keep the bird shut up in a cage and then blame the delicate little creature for not being able to fly. Under economic rule, I would moietise the "vote and influence" ticket solicited by our politicians, giving the votes to the men and all the influence to the women; and if men dared to exercise any influence, I would deprive them of their votes and enfranchise the women. If our politicians were *men*, no woman would ever dream of soliciting the franchise.

4. *Master and Servant.*—Those who have concluded that Humanitism makes provision for a dualism in reference to this question have not reasoned logically. When we once know what scientific wealth is, no fallacies can arise. The final increment of wealth is an increase in our physical development and our ethic attributes. All wealth commodities must tend to this end. From this standpoint every individual is his own capital, invested in his own business, and he keeps accounts with himself. His account-books therefore show the progress which he is making in his business. Every man must therefore be his own master, and if he can exchange services with his neighbour, no relation of master and servant can arise, neither can the actions or transactions be regarded as egoistic or altruistic. When an action is just, all the other qualities and attributes, if it can have any, must follow as a logical necessity. Master and servant is suggestive of a criminal relationship. The standpoint of development is that each individual should perform such actions, thus developing such functions, as are necessary for procuring the means of subsistence.

This basis should never be discarded, and it pertains to the whole animal kingdom ; but by no means can the inference be drawn that this sphere of activity should not be enlarged. Indeed, this enlargement should be the line of demarcation between man and the lower animals. In this manner, exchanges of services and transactions may be made, which can never lead to the relationship of master and servant. When an individual coerces a weaker brother to perform actions which he should perform himself, the inevitable condition is that the strong man grows weak and the weak man strong, the servant thus becoming master, and the repetition of this incident is the summary of all human history. Mankind conceived in the relationship of master and servant is the fabrication of a diseased brain, a pure nonentity, and can never have a concrete existence. If these words are to be retained in our language, we may all regard ourselves as masters and also as servants. If it be said that there are great men whom we must acknowledge as masters, then we strike another abstraction, for the inference must be drawn that weak individuals can be permanently perpetuated. If 'we acknowledge masters in one sphere of activity, we must be acknowledged as masters in another sphere. On what principle can we otherwise exist? If our race is to be perpetuated, we must gravitate towards equality, and when everybody is great, nobody is great. From the humanitistic standpoint, the number of great men are miserably few. Upon what principle can a man be called great who delights in war—in the shedding of human blood? How can a great man spend precious time in developing abstract speculations and building castles in the air? Can a great man delight in plundering his fellow-men, and in transmitting the spoils to his "heirs and assigns for ever"? Is he great who coerces his weaker brother to become his slave? The truly great man desires that every individual should have equal opportunities for becoming great and good, and bestirs himself to attain the object of his desire. He who employs a servant, as he would a brute, to administer to his abstract desires must abandon all hope of future rewards. As before remarked, however, such conditions must be fulfilled in the onward march of universal evolution; and from this standpoint the master, as a martyr in the cause of mankind, deserves our deepest sympathy. On the same principle, the priests in all ages should be deified for their martyrdom in the cause of humanity, for they have amply proved the fallacy of the theologic idea.

5. *Education and Instinct.*—It is admitted on all hands that we are artificial beings. If this assertion has any meaning, it must signify that we have sacrificed our instincts on the altar of education. This question may be discussed from two standpoints : (1) the material, and (2) the spiritual. All the lower animals are instinctively drawn into relation with the source of subsistence; but with man this is not a respectable employment. Being an educated animal, he attempts to live on luxuries and on the theories of abstract ideas, which tend to extinguish his instincts, and consequently also his very existence. In the spiritual world, it is quite respectable to theorise abstractly about the nature and attributes of virtue and abstract religion, but to speak about their true origin is a criminal offence. If we wish to reverence posterity, we must elevate and purify its source. Nothing should be purer than the source of subsistence and the receptacles of posterity. Education can only have one definition, namely, the misuse and abuse of our instincts, so that by confining our efforts to the cultivation of our instincts, the dualism is solved, and this is the only method of solution. That man can evolve into an educated animal is just as great a fallacy as the theory of his evolution into a theist or an orthodox theologian. We can only evolve by cultivating or educating our instincts; we certainly cannot do so through actions which tend to destroy them. The generalised ideas of the past are an objective deity—the product of our intellect; the generalised conceptions of the future are a subjective deity who presides over humanity, and is the product of our virtue, not of our intelligence—the product of instinct, not of education. The salvation of our race through education has proved a disastrous failure; the only recourse is to try the effects of our instincts. By this shift of basis we do not debase our intellect, but rather strengthen and purify it, for

Y

we then acquire the natural incentive to exercise our intellectual faculties, which, in their turn, strengthen and ennoble
our virtue.   In like manner, the shifting of the material
basis from the commercial to the agricultural will increase
and purify our manufacturing and commercial industries, for
a hundred millions of people will have more wants, although
no abstract desires, than a third of this number, and the
immense importations and exportations of human beings in
pursuit of physical and intellectual training will be much
more respectable and profitable than an increased trade in
swine and sausage.   But this gigantic manufacturing enterprise in humanitistic industrialism should not be driven to
its objective extremes.   By all means, let us have Professor
Huxley's "wilderness of apes," and a suitable number of all
other kinds of animals.   We have already learned more from
the ape than from all our educational institutions combined.
Professor Ape teaches us no abstract theories.   Judging from
the cruelties which economic men inflict upon, the lower
animals, there will not even be an ape to whom they can
hand over their manufacture and trade when man perishes
from the face of the earth, which must be his destiny under
economic rule.   Humanitism, by its very nature, inspires
reverence for the "dumb brutes," and no humanitist can
tolerate the merciless cruelties of the slaughter-house.   The
theory that God created the lower animals for man's use and
pleasure, which has led to the cruelties of the chase and the
slaughter-house, has no place in humanitistic ethics.   Any
change of categories in the ethical must lead to a corresponding shift of basis in the material world.   The forest is man's
home ; the city is his exile.   Trees are a condition of his
existence, and he can never evolve into a city animal.   Our
education drives us into the city ; our instinct drives us into
the forest.   The city does not bring human beings closer
together; on the contrary, it severs them farther asunder.
Material proximity is spiritual diffusion. The humanitist brings
physical, intellectual, and spiritual education into unity and
harmony, thus combining science and art, and dispenses
with practice.   Art signifies the carrying out of scientific
principles, so that practice can only signify the repetition of

acts which have no scientific basis. The word practice is, however, useful to express operations conducted in experiments. He does not ignore the past, as the heroes of the French Revolution have done, but rather stimulates historical inquiry, gathering from the pages of our ancestors and of nature all that will conduce to the fulfilment of our destiny. Man has no rights beyond those derived from the immutable and inexorable law; but he has a sovereign duty to perform, and that duty is to be obedient and just.

6. *Nationalism and Internationalism.*—Patriotism' should be sharply defined in every system of ethics. Granted that it is the duty of every loyal citizen to shoulder arms in defence of his country, yet the question still arises as to the nature of his patriotic duties where there is no monopoly, no over-population, and consequently no national or international wars. Under economic rule, it is the duty of the State to defend monopoly, and the more successfully it performs its duty, the greater, one would think, should be the patriotism of the subjects. But the duty of all civilised States is identical, although there may be variations in the performance, and it is therefore difficult to understand how a subject can be more loyal to the land of his nativity or adoption when the rulers perform their duty less efficiently than those of other States. Humanitism, in its name and essence, is inconsistent with nationalism; and an international boundary-line is a pure abstraction. And yet, despite this fact, it is the only *ism* which is workable in any given State without regard to the *isms* adopted in the adjacent or surrounding States, or even in distantly located States. For example, what could any foreign State do against the United Kingdom with a population of a hundred million of peaceful souls, all being supported from our own soil, every man, woman, and child being virtually a trained soldier, and there being no cities, towns, or villages to be attacked? The flimsiest conception of an invasion is sufficient to prove its absurdity; and if other States adopted our weapons of defence, there would be no aggressor, and consequently all warfare must come to a pause. Those of us who imagine that, under economic rule, war will be abandoned for the reason that nobody desires to see such

a disaster, or on account of the "spread of Christianity,"
are, to express it mildly, very illogical reasoners. If all the
nations of the world entered into a holy alliance for the
prevention of war and for the cessation of arms, the document would only be a signal of still greater disaster. War
waged for the promotion of peace is not less disastrous
than that waged to satiate the ambition of the greatest
despot who ever usurped or encumbered a throne. The
reason for this is, that economic forces are stronger and
more damning than the voices of the preachers of right-
eousness and peace. The proposal to change individual
into State monopoly still leaves these forces in the same
category. France by her Revolution tried this change,
and after a century's experience it has proved itself to be a
woeful failure. Monopoly is the parent of all disaster, and it
makes no difference if the monopoly is wrung from humanity
by the State or by the individual. When we learn how to
produce wealth without making it scarce, and when we see
that heaven, with its deity, is in the future and not in the
past, we inaugurate a period of universal and everlasting
peace. When we speak of coming wars as being engines of
disaster amongst nations, we display our ignorance : they are
wars between theism and metaphysics—struggles between
supposed supernatural powers external to an object and those
inherent in it. It makes no difference whether this supernatural agency comes down to us from the shades of our
ancestors, or whether it is inherent in gold or in the crown.
None of these agencies can save us or point us out the way to
salvation. The spread of theism would drive us back to the
Oriental Theocracies of three thousand years ago ; an orthodox
priestcraft combination would restore us to the Middle Ages ;
and the metaphysical idea would plunge us into the year 1789,
thus inaugurating a universal era of French fratricidal Fraternity. There are two conflicting views entertained with reference to the French Revolution : (1) that it has been a grand
success, and (2) that it has been a disastrous failure. Both of
these views are right, and both are wrong. All depends upon
the set of theories from which the facts have been derived.
From the standpoint of universal evolution, the French Revolu-

tion must be pronounced the most brilliant achievement in the history of mankind. It has proved that the metaphysical is just as chimerical as the theological idea. On the other hand, from the standpoint of its default in attaining the end at which it aimed, it must be pronounced a most disastrous failure. It aimed at the solution of the economic problem; but instead of solving it, the horrors of economicism have been brought out in all their intensity. A vigorous era of Socialism would be death to theistic and theologic abstractions, and would add fresh evidence to the inefficiency of the metaphysical idea; but the economic question is still as backward as it was a century ago. We in England are not far from the time when a Trafalgar Square mob can transmute us into a Republic; and then, as in France, begins the era of dictators and despots. The civilised States of modern times do not depend upon foreign barbarians for their overthrow: modern barbarians are bred in streets and garrets, and Christian civilisation is their mother. Foreign barbarians are being brought under the yoke of civilisation by our Bible and our beer. What hope can there be for the future of our race so long as Oriental Shahs continue to come West to take models of our Western civilisations? Our greatest foe is indifference, and we may continue to boast of our flag and our commerce until we behold the naked fishermen swarming to line our native shores. Those of our rulers who persuade themselves that the soldiery, so long as they are drawn from the masses, will much longer continue to obey the fiat of merciless despots must suffer the consequences of their miscalculation. Our critics flatter themselves that their task is done when they point out the errors and inconsistencies of proposed changes, or prove them to be worse than our existing conditions. The real question is no longer confined to the impracticability of proposed institutions— indeed, it is safe to infer that all changes in the abstract category will be for the worse—but our present concern is how we are to conduct ourselves in the inevitable revolutions of the near future. So long as we can find an outlet for our poverty-stricken and rebellious masses we may avert the disaster, but our time should be employed more profitably and

manfully. Warning cannot come too soon; it may come too
late. Peace is growing more terrible than war. To say that
man is by nature a war animal is to enunciate a theory which
has no foundation in fact, and the assertion is a very convenient method of palliating our crimes. I would regard
myself guilty of a great wrong if I made such an accusation
even against carnivorous animals. Man, like all other animals,
has strong self-preservative propensities; but to say that he
enjoys the scent of human blood is a vilification of his instincts. As an educated animal, he may exult in the gore of
his prey, as the carnivores do, but to rank his natural com-
passions below those of other herbivorous animals is the inflic-
tion of unscrupulous injustice.

I have now offered a solution for six dualisms which have
been the source of much discord in all ages. All abstract
speculations have failed to approach unity and harmony in
these questions; indeed, the discordance is growing more and
more strained. There are other questions which bear upon
two or more of these dualisms, one of which I will now briefly
notice, namely, language.

Language is one of the best measures of man's progress or
degeneracy. It is tacitly assumed by some that the nose is the
organ of speech, by others the tongue, by others the mouth.
If each structure is to exercise its natural function, this question must be solved. All these assumptions are false, and are
incontestable proofs of our degeneracy. Just on the same
principle which makes us too lazy to exercise our jaws to
masticate our food, thus attempting to cast the burden on
organs which have not been constructed to bear them, we find
a tendency to talk with organs which are not intended for
speech. When we evoke sounds from the chest, we produce
speech and melody; when we empty other organs, we talk
and hum. The recital of one of Homer's choice hexa-
meters is sufficient to exhaust our vocal resources. That
the ancient Greeks possessed great physical strength is
proved by their language, as well as by the statuary
which has been exhumed from their ruins. They *spoke*
their charming and powerful language; they did not *talk* it,
as we do ours. This is proved by the deep and full sounds of

the vowels and diphthongs, and the absence of silent consonants. When all the letters in all the words of a language become silent, there is a sure mark of physical and moral degeneracy. The leading characteristics of a language may be divided into three : (1) deep, full vowel sounds, (2) inflections and terminations, and (3) the absence of silent letters. The beauty of English, therefore, is that it is not a language at all. Full-sounding vowels must be pronounced by the proper organ of speech ; inflections give variety of construction, and when all the letters are pronounced, the language can be spelled. The only language of modern Europe is the German, and Germany is the only country which, other conditions being equal, can produce orators and musicians. We waste in the spelling-class the time which the German spends in becoming a soldier, and the physical strength thus acquired is conducive to the production of a powerful language. We have no chest, and can therefore have no language. We have abundance of fresh air which is capable of producing the most powerful and bewitching language that has ever issued from a throat ; but owing to the very abundance of our air, we merely breathe to live : when it becomes scarce, we will live to breathe, and then, unless we go to excess, our talk will develop into speech, and our buzz into music. The Latin language was less perfect than the Greek, but more perfect than ours is, or any of the Romance languages, so that the progress of degeneracy may be easily traced. These preliminary observations are necessary to the main issue. Our so-called language is largely composed of Greek and Latin roots ; otherwise expressed, we have been obliged to steal the words, and have denuded them of their power and charm. What greater proof of degeneracy can be desired ? But this is not yet half of the whole story. We have also been obliged to steal the institutions and laws which were originally expressed in these heathen languages ; then we boast that our institutions are divine, and call ourselves *British* patriots. Our crown virtually decked the head of Aristotle, and this is the main source of our loyalty and our patriotism. To be ignorant of these facts is called education—classical education. But our story yet remains half untold. Our art is improved, and our virtue strengthened, by sketching the nude Gods and Goddesses of heathen Greece, and it is only from this source that our ideals of perfection in the forms of human beings can be obtained. One of the virtues which I claim for Humanitism is, that our harangue will develop into speech upon scientific principles, and not, as now, be left to the fortuitous concourse of circumstances; and the same remarks will apply to our progress in literature, art, and politics.

The necessity for a Deity may be questioned. It may be urged that humanitistic forces will drive people into morality, Deity or no Deity. This statement is quite true; but it must not be forgotten that we cannot *individualise* the conceptions of posterity; we must *generalise* them, and God is a symbol of unity, order, and harmony, leading to concentration of thought and feeling, so that we can most effectually fulfil our destiny. Mellos is a patron of art and strengthens our subjectivity. Besides, I am aware of no object in life except happiness and pleasure, and it seems to me that art, founded upon science, should be the orthodox method of attaining this end. The condition of our art is just as deplorable as that of economic and theologic orthodoxy, and of law and politics, the reason being that our artists have, during all these long ages, been working without a vanishing point. Mellos is the vanishing point of all our feelings, thoughts, and actions. If it be an essential characteristic that a Deity should be invisible, uncalculable, and unknowable, these qualities being necessary for the inspiration of awe and dread in his worshippers, then I must confess that Mellos possesses a weak spot—a vulnerable heel. He possesses this great advantage, however, that he is very deceptive; for the perfection of art consists in its deception. The artist, for example, who cannot depict a cluster of grapes so as to make our teeth water, or a sausage that excites our contempt and loathing, is a mere novice in his calling; and the sculptor who cannot chisel life into his divine image belongs to the same category. An unknowable and unrevealable Deity would have no scientific or artistic worth, and would quench our noblest aspirations.

I have no sympathy with those philosophers who insist that happiness can exist without pain. We might as well attempt to conceive of light without darkness. The tenderer the love the mother bears for her once darling child, the greater the tears she sheds over its grave. He who is not acutely pained by the distressing scenes of vice and misery which he constantly beholds all around him is dead to all pleasurable sensations. The virtue of pleasure consists in its freeing us from the pernicious effects of inordinate pains. Stupefactive pleasures, the harbingers of pain, are the logical necessity of abstract methods of thinking. Would the sum of human pleasures have been diminished had voluptuous indulgences never existed? To scientific pleasures there is no limit; they are appropriate at all times and in all places; but those begotten of luxury and indolence have a very narrow range. Had soporific pleasures no existence, we could have no conceptions of them, and our thoughts and feelings would be moulded for the enjoyment of natural pleasures. Consider, too, the egoistic nature of sordid indulgences; for the displeasure in foregoing the satiation of our debased appetites is stronger than our thoughts of the pain which we inflict upon those who are solicitous about our welfare. Natural pleasures are bound up with utility. No pleasure can be derived from an action which conflicts with our moral destiny, so that happiness may thus be brought under mathematical rules. Happiness is derived from seeing a structure perform its characteristic function; the degree of pain may be measured (1) by the neglect of such performance, and (2) by the structure performing a false function. The pain is in exact ratio to the non-performance, or to the falsity of the performance. Utility is also allied with beauty. There can be no beauty without utility, and the beauty of an object is in direct ratio to its utility. Nothing can be beautiful which engenders inordinate pain, and all inutilities are sources of pain. Beauty cannot be said to reside in activity or repose. When a structure has exercised its function to its fullest capacity without strain, beauty consists in repose; when the Goddess of repose has her due, beauty abides in activity. It is beautiful to weep when we behold an attitude of pain; it is beautiful to laugh when we behold actions or objects which are suggestive of pleasure. The more we cultivate happiness, the keener grows our sensibility to pain. The more we deaden our sensibility to pain, the blunter grows our susceptibility for happiness.

The humanitistic conception of truth is admirably defined by George Henry Lewes,[1] namely: "Truth is the correspondence between the order of ideas and the order of phenomena, so that the one becomes a reflection of the other—the movement of Thought following the movement of Things." The beauty and force of this definition consists in this, that the deviation of the movement of the thought from the movement of the thing is a measure of man's free-will—his powers of objectivity. In the abstract method of thinking, an attempt is made to force the thing to follow the movement of thought. It is important to bear this in mind, because, as I have shown, we are governed by abstract methods, and the only recourse is to reverse all our activities and habits of thought before a solution of the social problem can be expected. In the ultimate analysis, justice consists in granting to each individual in all generations equal opportunities for access to the means of subsistence. All other conceptions of justice may be logically derived from this truth. Light is truth diffused, the access to which each should also have equal opportunities: hence the corner-stones of Humanitism are Truth, Justice, and Light. Even these abstractions, however, must ultimately vanish.

In speaking of the forces by which we are governed, I have suggested that they should be shifted from the abstract to the scientific category. As we have seen, there are two sub-categories: (1) the material, and (2) the spiritual. If we now admit, and act upon the admission, that trade should be created for the good of man, and not man for the good of trade, we find ourselves in the scientific material category. Again, if we admit, and act upon the admission, that heaven with its divinities is in the future, and not in the past, we find ourselves in the scientific spiritual category. We cannot shift the material without the spiritual, or the spiritual with-

[1] History of Philosophy, vol. i., xxxi.

out the material; otherwise we beget a dualism which must end in the most terrible of all human disasters. In the scientific category, the Humanitist is quite consistent in assuming the following titles: 1. Materialist. 2. Spiritualist. 3. Individualist. 4. Socialist. 5. Positivist. 6. Agnostic. 7. Secularist. 8. Rationalist. 9. Utilitarian. 10. Humanitarian. 11. Theist. 12. Christian. This simply proves that all dualisms are merely abstract conceptions, which have no foundation in nature; but it is necessary to make some remarks upon the claims of the Humanitist to be called a Christian. His ultimate claim rests upon the assumption that Christ was good and sane, and therefore also consistent. Ignore this assumption, and the Humanitist is not a Christian. Now, as Christ wrote nothing Himself, and as there are conflicting accounts of what He said and did, all the contradictions which are inconsistent with goodness and sanity must be the work of priests, scribes, or other interpolators. The divinity of Christ need not be denied, for His inspiration appears to be quite consistent with human aspirations and instincts. Christ being a good man, He must have been inspired by a good God who could not be so wicked as to damn human souls, which are part and parcel of Himself, to be eternally lost in a lake of fire. If Satan be responsible for this torture, then it was a wicked act for God to create Satan, and if He did not create him, he must be a person in the Godhead, and therefore God is good and bad, and is not omnipotent. To be a devil, and to create one, are equally iniquitous. As Christ could only have preached the goodness of God, the priests must have been the fabricators of the Plan of Salvation. Some modern theologians attempt to evade the wicked character of the Godhead by saying that everlasting fire does not mean a fire lasting for ever, according to the modern translation of the Greek word which conveyed the terrible idea. If this be true, then eternal bliss cannot last for ever, and the New Testament cannot be a revelation from God; for a revelation must reveal, and cannot therefore confuse. A divine revelation confused by priests cannot be an inspiration from an omnipotent, omniscient, omnipresent, and a just and holy God. Another claim of the Humanitist is, that Christ says *Come;* He does not say *Go*. He invited human souls to *come* to the *future*, not to *go* to the *past*. While claiming to be a Christian, the Humanitist repudiates priestcraft Christianity, and pleads that Christ possessed the true subjective spirit of religion.

Humanitism is specially to be recommended on the ground that it saves alike the unemployed rich and the unemployed poor from perdition. Its supreme aim is not to destroy property, but merely lessen pride and folly. A rich man suffers a loss of property only when his neighbour gets the benefit; but if the individual loses his surplus property, his pride and folly being reduced to the same level, no actual loss can be sustained; contrariwise, all is gain. Humanitism does not direct its missiles against property, but primarily against pride, ignorance, and folly, and against the sovereignty of the Idea. We are all the creatures of circumstances, and are therefore deserving of compassion. We would all be good if we could; but we are coerced to do those things which are revolting to our nature. We are brought up within a narrow compass, and our self-preservative instincts being strong, we would rather bear those ills we have than fly to others we know not of. The greater the future dangers the more tenaciously we cling to what we vainly call our own ; and the less able we are to obtain a livelihood the more we cling to our property. I am therefore convinced that every proposed reform should be accompanied by rational methods for carrying it out. All healthy changes should be imperceptible. While recognising these facts, I present this portion of Humanitism as being the plans and specifications; and as for the executive powers, it is my purpose to show, on a future occasion, that such matters do not belong to mere men of affairs, but enter into the construction of a special science. No single scheme can efficiently carry out any great reform; and I am prepared with a combination of schemes, should no flaws be found in my plans and specifications, and should a demand arise for their execution. I have not lost so much faith in my fellows as to assume that they desire the present state of society to continue, providing a rational method of escape from existing ills be found, and it has therefore been my aim to present as clearly, yet briefly, as possible a new condition into which no fair-minded person need be afraid or ashamed to enter. Although I have on both hemispheres beheld with my own eyes many heart-rending scenes of misery, and listened to many a distressful tale of woe with my own ears, and although I am not unfamiliar with painful statistics gathered from other sources, yet the violent character of our social, political, and economic dualisms is in itself sufficient to convince me that civilised society is in a state of anarchy, and that there is no possibility of solving the social problem through any one or more of the methods which have hitherto been proposed. The entire basis of our social forces must be changed. Although my arguments may not receive general acceptance, yet they cannot fail to show the limited nature of the sphere in which the champions of the various schools of sociology are operating. I may have been rather polemic in some of my pages, which may appear inconsistent with a purely scientific inquiry, but I cannot forget my duties and responsibilities as a social reformer. My supreme aim, however, has been an inspired regard for truth ; and if I have given needless offence, it has arisen from the fact that those whom I have criticised appear to have acted under indefensible motives. I do not maintain that economicism, politics, and religious orthodoxy have been barren of results. They have taught us many an invaluable lesson, and have been the means of forming a bond, more or less friendly, between nations, giving them opportunities for becoming more intimately acquainted with each other's language, literature, and laws, which bond is a colossal stride in the cause of humanity. The rapid progress which the doctrine of evolution has made amongst the acutest minds in France and Germany, as well as amongst those in our own country, is another step in the same direction. Science and mathematics have been brought into disgrace by the attempts made through them to justify abstract theories, and the reaction due to this failure must have an important bearing in relation to the social problem. Agricultural science has demonstrated that the soil is not an inexhaustible mine, and that the enormous waste caused by artificial methods of husbandry must be checked. There is great scope for development in agriculture, which is the sum of all the sciences, but so long as it remains a slavish and degrading occupation, abstract pursuits must continue. All the sciences should be taught in their relation to agriculture. With reference to the political situation in all civilised countries, the potentates, whose thrones are kegs of fizzing dynamite, are being tremblingly weighed in the political balance, or are fugitives to lands where they can find no security for their bodies or rest for their weary souls. There can only be two sides in the coming conflict, namely, ancestry against posterity, the former being the aggressor, and the weakness of the latter is the source of its strength. Each must choose for himself, whether it is his duty to take up arms against his children or against his ancestors. There is no halting-place between these extremes. We must go to futurity or be gathered amongst the ashes and shades of our ancestors; we cannot stand still. We may continue the struggle a short time longer, but if we do, it must end in the fall of our once cherished civilisation. I have not entered into my inquiry in the spirit of a national destroyer, but rather with a patriotism only shadowed o'er by my affection for humanity, and I hope that my effort will be criticised with the spirit in which it has been written, and with the sense of truth and justice which the subject imperatively demands.

THE END.

PRINTED BY BALLANTYNE, HANSON AND CO.
EDINBURGH AND LONDON.